Mechanical Desktop 6
Visual Fast Start

Craig Stinchcomb

Upper Saddle River, New Jersey
Columbus, Ohio

Library of Congress Cataloging-in-Publication Data

Stinchcomb, Craig.
    Mechanical Desktop 6 : visual fast start / Craig Stinchcomb.
    p. cm.
Includes index.
ISBN 0-13-096902-8
1. Engineering graphics. 2. Mechanical desktop. 3. Engineering
design--Data processing. I. Title.
T353 .S819 2003
620'.0042'02855369--dc21

2002005456

Editor in Chief: Stephen Helba
Executive Editor: Debbie Yarnell
Editorial Assistant: Sam Goffinet
Media Development Editor: Michelle Churma
Production Editor: Louise N. Sette
Production Supervision: Karen Fortgang, bookworks
Design Coordinator: Diane Ernsberger
Text Designer: STELLARViSIONs
Cover Designer: Tom Mack
Production Manager: Brian Fox
Marketing Manager: Jimmy Stephens

This book was set in Berkeley by STELLARViSIONs. It was printed and bound by R. R. Donnelley & Sons Company. The cover was printed by Phoenix Color Corp.

Warning and Disclaimer: This book is designed to provide tutorial information about the Mechanical Desktop computer program. Every effort has been made to make this book complete and as accurate as possible. But no warranty or fitness is implied.
    The information is provided on an "as-is" basis. The author and Pearson Education, Inc., shall have neither liability nor responsibility to any person or entity with respect to any loss or damage in connection with or arising from the information contained in this book.

Pearson Education Ltd.
Pearson Education Australia Pty. Limited
Pearson Education Singapore Pte. Ltd.
Pearson Education North Asia Ltd.
Pearson Education Canada, Ltd.
Pearson Educación de Mexico, S. A. de C.V.
Pearson Education—Japan
Pearson Education Malaysia Pte. Ltd.
Pearson Education, *Upper Saddle River, New Jersey*

Copyright © 2003 by Pearson Education, Inc., Upper Saddle River, New Jersey 07458. All rights reserved. Printed in the United States of America. This publication is protected by Copyright and permission should be obtained from the publisher prior to any prohibited reproduction, storage in a retrieval system, or transmission in any form or by any means, electronic, mechanical, photocopying, recording, or likewise. For information regarding permission(s), write to: Rights and Permissions Department.

10 9 8 7 6 5 4 3 2 1

ISBN: 0-13-096902-8

PREFACE

With today's incredible technological pace, even the most dedicated technologist has a constant feeling of "running uphill" when it comes to learning new software. There is so much to learn and so little time. We know we need to learn the new software, we know all the promises this new technology has to offer, but when can we possibly dedicate the necessary time? This is where *Mechanical Desktop 6: Visual Fast Start* comes in.

Mechanical Desktop 6: Visual Fast Start is exactly what the name implies, a highly visual, very fast method of learning. You will find that this text works like your own learning process, visually! As you read, your mind develops pictures. These pictures develop into concepts, and your mind then attempts to translate them into steps or processes. This process takes time and can make the learning process difficult and painfully slow. With *Mechanical Desktop 6: Visual Fast Start* we speed this process by going directly to the visual stages of learning. We provide the visual steps that are necessary to make learning complex topics easy and actually fun to learn. *Mechanical Desktop 6: Visual Fast Start* uses carefully laid out step-by-step processes and pictures of the concepts. You will not only learn faster, but will also absorb more details and get the full benefit the software has to offer.

Each chapter of *Mechanical Desktop 6: Visual Fast Start* has 3 parts:

- A brief process explanation and step-by-step tutorial
- A section entitled "For a deeper understanding about"
- "Real-world" practice drawing projects

The tutorials will give you the initial jolt of understanding, whereas the "For a deeper understanding about" section will satisfy that need for detail once the concept is mastered. The finishing touches are added as you practice the "real world" projects at the end of each chapter. Each of these sections is designed to make learning Mechanical Desktop fast, easy, and FUN! Whether you are a new student to Computer-Aided Design software, or a long-time veteran, *Mechanical Desktop 6: Visual Fast Start* will prove you really can conquer the new and complex design software.

ACKNOWLEDGMENTS

Special thanks are due to the "wizards" at Autodesk who continue to amaze us with better, faster, and easier to use Computer-Aided Design software tools. I would also like to acknowledge the reviewers of this book: Lou Moegenburg, University of Wisconsin, Stout Campus, and David Steinhauer, Tidewater Community College (VA).

Also, thanks to Jon, James, Reese, Aimee, and my wife Sally for all their patience and love.

CONTENTS

1 INTRODUCTION   1

Here Is Where It All Starts...   1
Basic 3D Modeling Terminology   3
Mechanical Desktop Display Modes   6
Your First Mechanical Desktop 3D Model   8

2 CONSTRAINING   13

The Basic Steps of Parametric Modeling   13
The Primary Steps in Constraining   14
For a Deeper Understanding about Constraining   22

3 3D MODELS: TURNING A CONSTRAINED SKETCH INTO A 3D FEATURE MODEL   27

For a Deeper Understanding about Extruding   29
Adding Holes, Fillets, and Chamfers to a 3D Feature Model   32
For a Deeper Understanding about Fillets and Chamfers   37

Viewing 3D Models   40
For a Deeper Understanding about Viewing Models   42
Handy Single-Key Viewing Shortcuts   44

4   CREATING A NEW PART   49

Changing Working Faces   50
Adding New Features to a Part   52
Editing the Model with the Desktop Browser   54
The Difference between Sketch Planes and Work Planes   57
Work Axes   58
Work Points   59
For a Deeper Understanding about Work Planes   59

5   REVOLVING   71

For a Deeper Understanding about Revolving   74

6   SWEEPING   83

For a Deeper Understanding about Sweeping   87
Lofting   88
Fixing Lofting Problems   92
For a Deeper Understanding about Lofting   93

7   THE SHELL COMMAND   103

For a Deeper Understanding about Using Shell   106

8   CREATING DETAIL DRAWINGS (BLUEPRINTS) FROM PARAMETRIC MODELS   109

For a Deeper Understanding about Detail Drawings (Blueprints)   116
Adding and Adjusting Dimensions on a Drawing   121
Advanced Orthographic Drawing Projections   128

9 HELIX   141

For a Deeper Understanding about Helix   147

10 PARAMETRICS   151

Creating Parameters in a Model   151
For a Deeper Understanding about Parametric Equations   158
Parametric Modeling with Table-Driven Variables   161

11 ASSEMBLIES   173

For a Deeper Understanding about Assemblies   186
Assemblies as Design Tools   189

12 ADVANCED ASSEMBLIES AND BILL OF MATERIALS   195

13 ADDING SYMBOLS TO A DRAWING   207

14 MECHANICAL DESKTOP POWER PACK 3D CONTENT: PREMODELED PARTS AND HARDWARE COMPONENTS   217

For a Deeper Understanding about Using Premodeled Parts and Hardware Components   223

15 FINITE ELEMENT ANALYSIS   231

For a Deeper Understanding about Part Analysis   236
Obtaining the Solid Mass Properties for Parts   237

16 USING THE MECHANICAL DESKTOP INTERNET TOOLS   243

TOOLBAR QUICK FINDER   253

INDEX   255

WHAT'S ON THE CD   263

INTRODUCTION

HERE IS WHERE IT ALL STARTS...

When you first open Mechanical Desktop you will see the introductory screen, which presents you with three major options for starting:

1. Open Drawings
2. Create Drawings
3. Symbol Libraries

Choosing **Open Drawings** will display a list of recently created drawing models from which you can select. Choosing **Browse** allows you to find drawing models you have previously created.

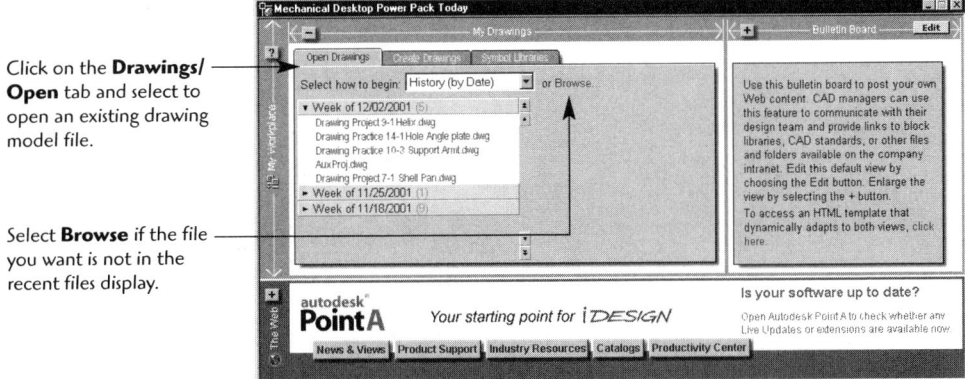

Click on the **Drawings/Open** tab and select to open an existing drawing model file.

Select **Browse** if the file you want is not in the recent files display.

Choosing **Create Drawings** will display the startup options **Templates, Wizards,** or **Start from Scratch**.

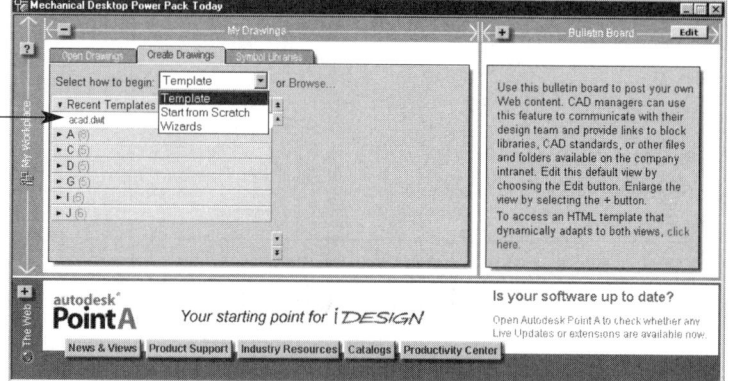

If in doubt about which template to use, select **acad.dwt**. This template is the standard blank 12 x 9 unit screen size. The Units and Limits can be reset to suit any size part later.

- The **Template** option will load a drawing *template* such as an ANSI size template (the .DWT file extension indicates a template file).
- The **Start from Scratch** option will ask you to choose *English* or *Metric* units of measure before starting the Mechanical Desktop Modeling screen.
- The **Wizards** option will walk you through a step-by-step setup of your drawing. You can select *Quick Setup*, which will simply have you input the units and the area of your drawing, or *Advanced Setup*, which will have you enter units, angles, and area options.

Choosing the **Symbol Libraries** will display an array of symbols and part shapes that can be used to detail your drawing.

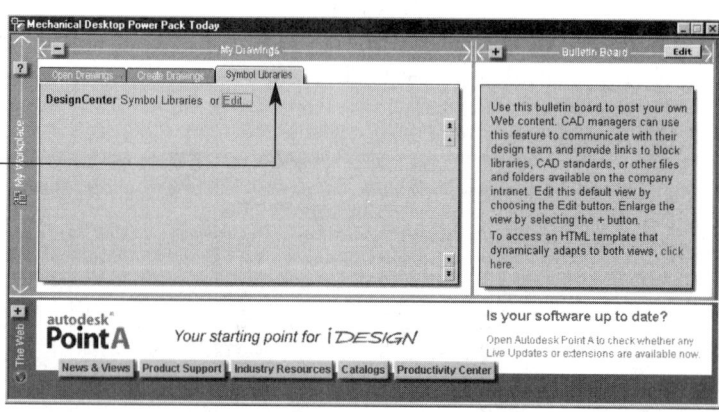

Selecting the **Symbol Libraries** tab opens the **DesignCenter,** which contains many 2D shapes and predrawn symbols. These can be used in the orthographic drawing portion and detailing of your design work.

Selecting the **Open Drawings** tab will display some search aids to help you locate drawings and models you have previously worked on.

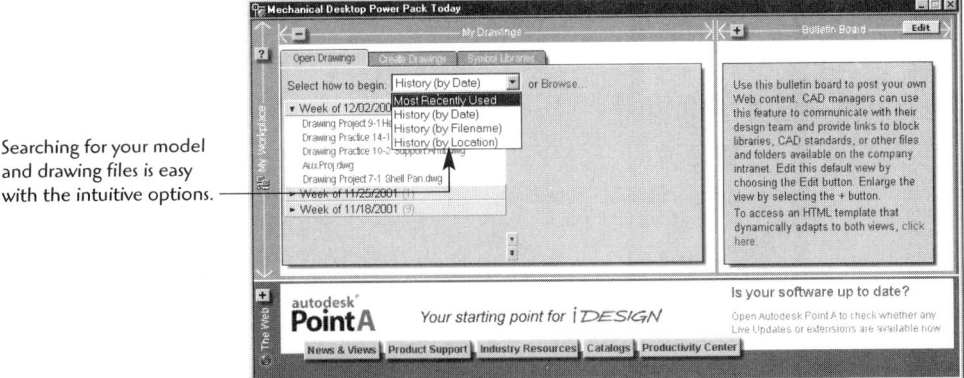

Searching for your model and drawing files is easy with the intuitive options.

BASIC 3D MODELING TERMINOLOGY

You will note that we will use the terms *model*, *drawing*, and *assembly* as we work our way through Mechanical Desktop. Let's clarify the differences:

A 3D **wireframe model** contains fully defined X-, Y-, and Z–axis dimensions.

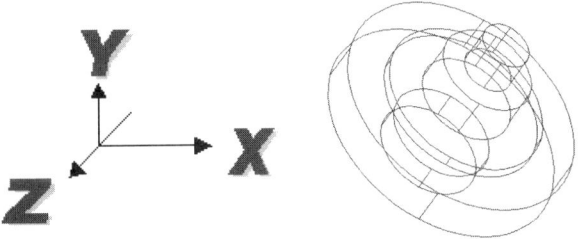

In addition to the properties of a 3D wireframe a **solid model** has *mass properties*. With the mass properties you can define the type of material such as aluminum or carbon steel.

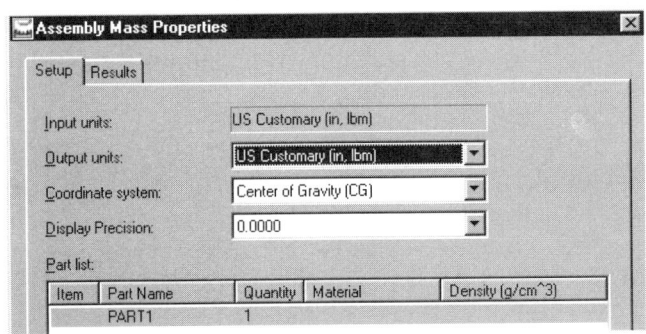
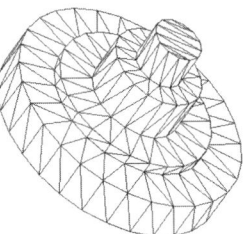

In addition to the properties of a solid model a **parametric model** has adjustable parameters. The adjustable parameters are established by constraining the model as it is created. Parametric models can easily be adjusted and modified to test different size configurations. This process is sometimes known as "what if" designing.

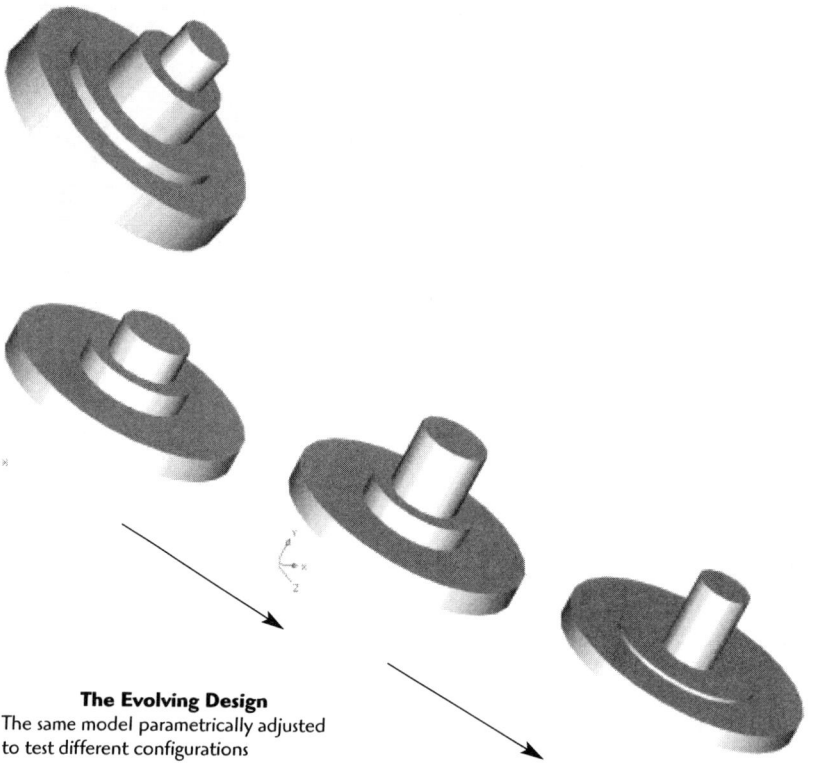

The Evolving Design
The same model parametrically adjusted to test different configurations

In addition to solid parametric Models, Mechanical Desktop allows you to produce *assemblies, to test the fit and check for interferences,* and *to generate highly detailed orthographic drawings.* The orthographic drawings can have isometric, assembly, detail, and section views as well as other standard orthographic model projections.

INTRODUCTION | 5

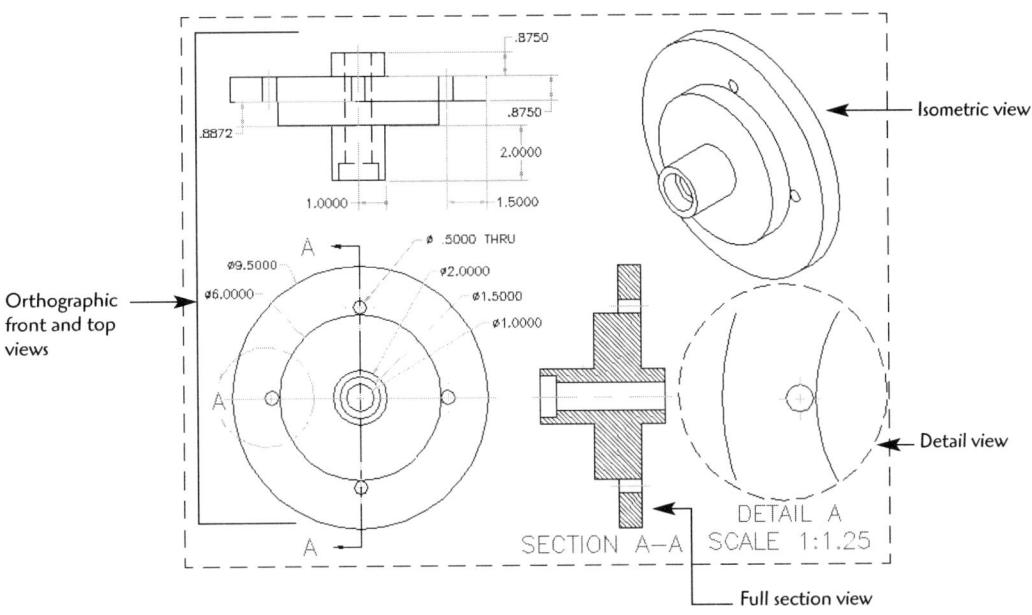

Orthographic front and top views

Isometric view

Detail view

Full section view

In the scene mode you can generate a number of scenes of your assemblies.

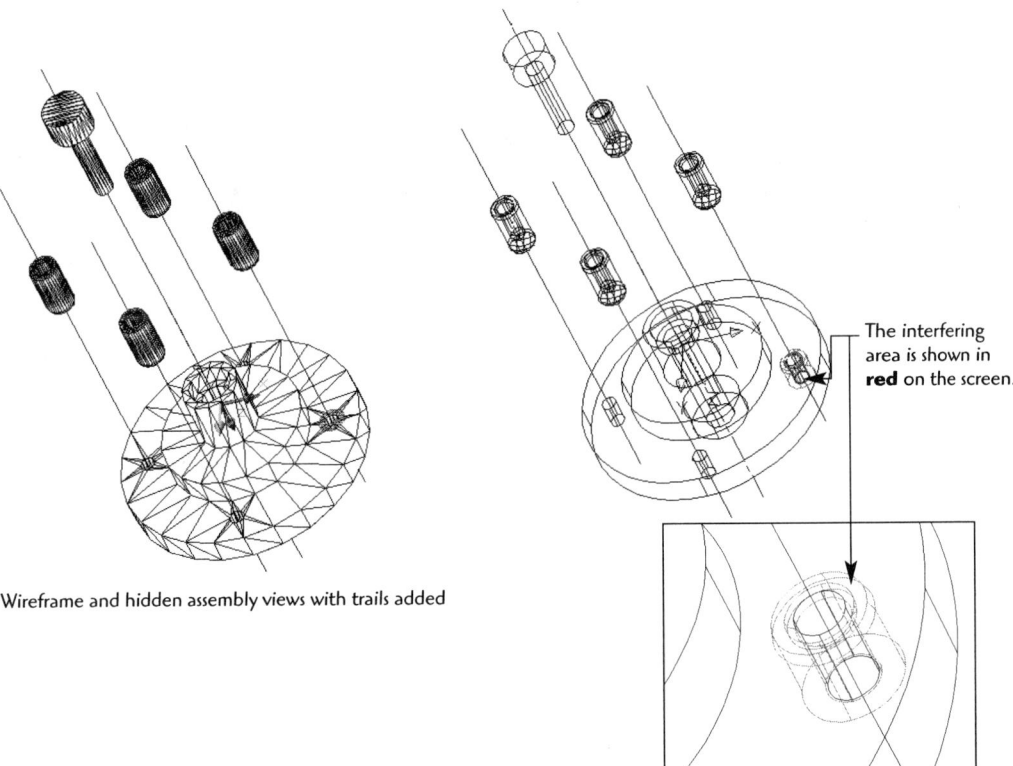

Wireframe and hidden assembly views with trails added

The interfering area is shown in **red** on the screen.

MECHANICAL DESKTOP DISPLAY MODES

There are three modes of display that you will work with when generating and displaying your designs and drawings:

- **Model** (You develop and edit parts and assemblies in this mode.)
- **Scene** (You display assemblies and tweak the views in this mode.)
- **Drawing** (You project, edit, and dimension orthographic and isometric drawings in this mode.)

In the model mode you will develop, edit, and test your designs. This is also where the designing process starts. You usually start with a two-dimensional (2D) **profile**, and then extrude, revolve, loft, shell, or sweep the profile into a three-dimensional (3D) **feature**.

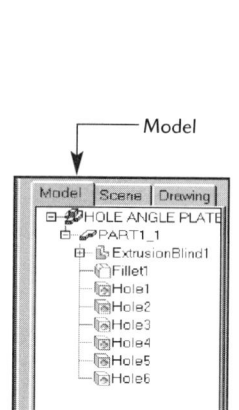

Model

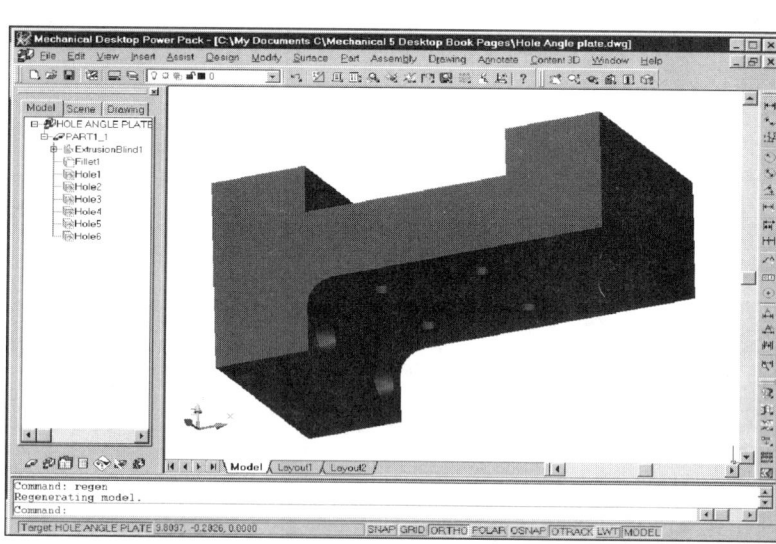

INTRODUCTION | 7

In the scene mode you display views such as exploded views and assemblies. The advantage of the scene mode is that assemblies can be displayed without any additional drawing. Scene mode lets you take the parts we have produced in an assembly and set an **explosion factor** to display the assembly. You can also include "tweaks" to adjust the view and "trails" to enhance and clarify the assembly.

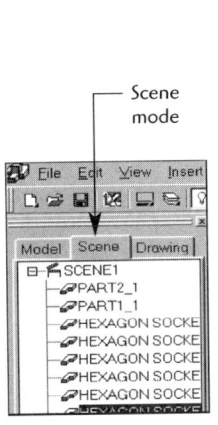

Scene mode

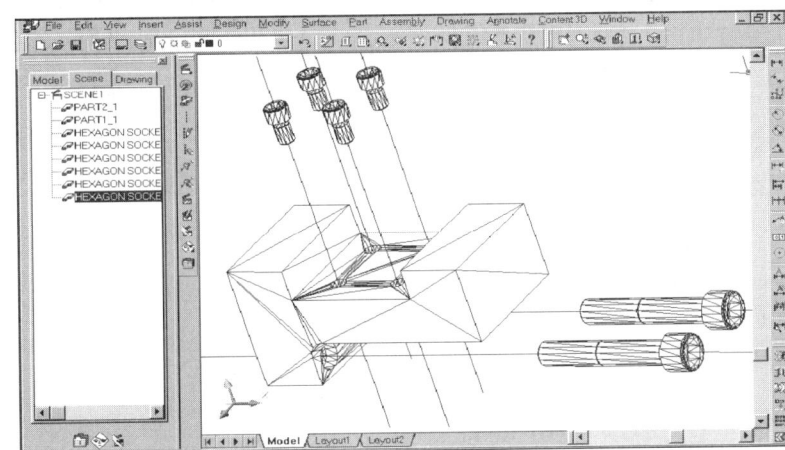

In the drawing mode you can display the conventional orthographic drawing views. Top, front, sides, sections, details, and other drawing views can be produced in this mode. Again, the advantage to this mode is that once the model or assembly is produced, the drawing views can be created from the model; no further drawing is required. This mode allows us to add additional dimensions and notes and generally to customize the display to suit your needs.

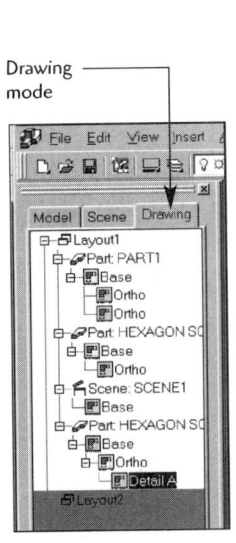

Drawing mode

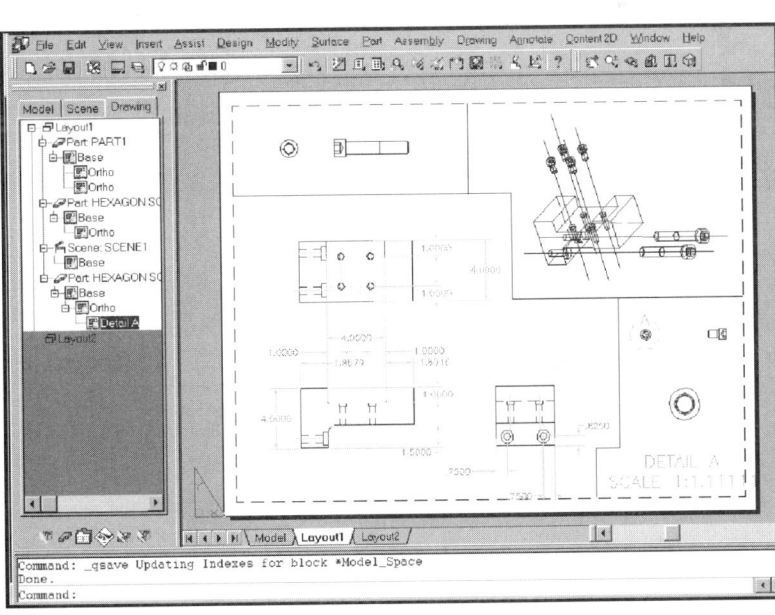

YOUR FIRST MECHANICAL DESKTOP 3D MODEL

Let's start with a demo drawing model to give you the feel of the Mechanical Desktop modeling process. Don't worry about why we use certain commands at this point. We'll cover that later.

Step 1   Start Mechanical Desktop, click on the **Create Drawings** tab, select **Start from Scratch**, and select **English** for the units.

Step 2   Type **PL** ↵
   or click the **Polyline** button.
   Sketch a basic L-shape with the polyline (the size is not critical now).
   Type **C** ↵ to close off the last line of the L-shape (the shape must be a closed profile).

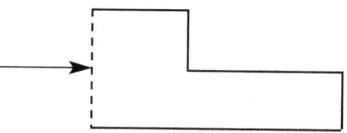

Step 3   Type **AMPROFILE** ↵
   or click the **Profile** icon.
   Select anywhere on the L-shape and press ↵.

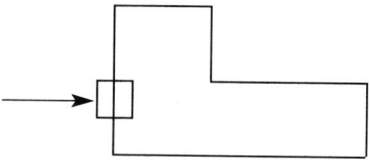

Step 4   Type **AMPARDIM** ↵
   or click the **Constrain Dimension** icon.
   Click on a line on the side of the L-shape, then click away from the line.

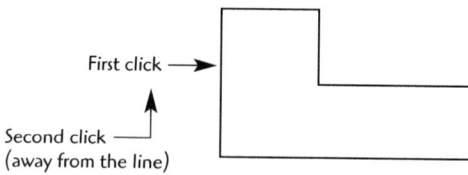

The command line will show a dimension. Press ↵ to accept the dimension.

This side is now "constrained." (We will explain constraining in detail in the next section.)

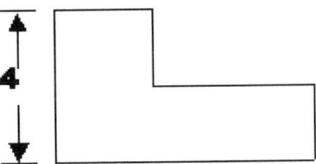

Continue around the L-shape until the message "Solved fully constrained sketch" appears on the command line.

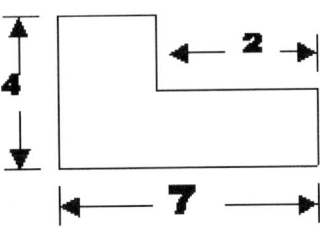

Step 5 Type "8" ↵. (The L-shape is now viewed on an isometric angle.)

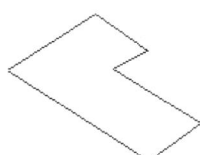

Type **AMEXTRUDE** ↵,
or click the **Extrude** icon.
Select the L-shape profile and press ↵.
Enter a distance of **1.0000**.
On the Termination option select **Blind**.
Click **OK**.

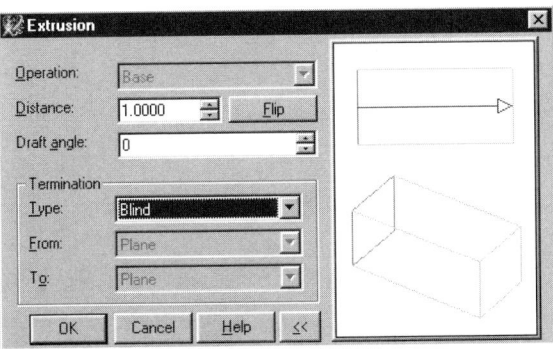

You now have the 3D feature, but let's add some details to the feature model.

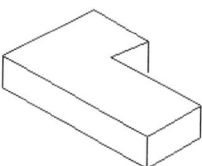

Step 6   Type **AMFILLET** ↵.
In the Fillet option box enter **Radius .5**
Leave the Constant box checked.
Click **OK**.
Pick the inside corner of the shape and press ↵.

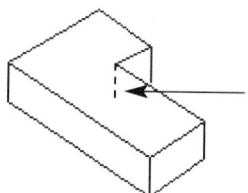

You can use the **Orbit** tool button to examine your first 3D model.

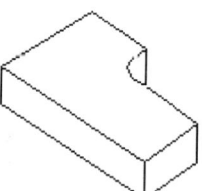

The completed model

Drawing Project 1–1

L-Block

Model the L-block to the exact dimensions indicated on the drawing. When you have finished, click **8** for an isometric view, and print out a copy for your instructor.

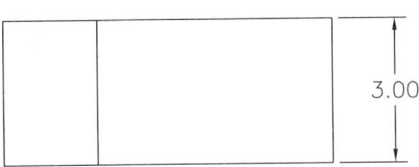
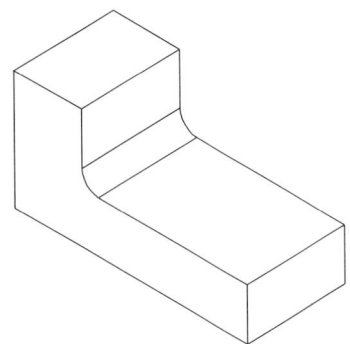
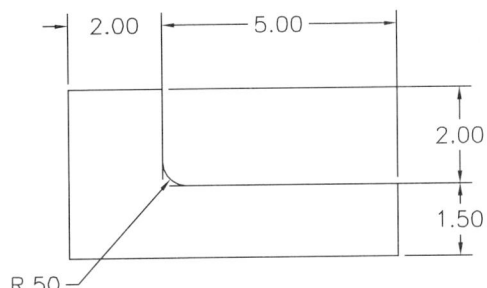
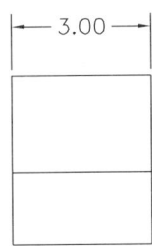

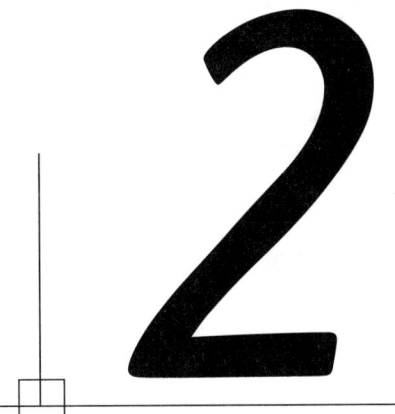

CONSTRAINING

THE BASICS STEPS OF PARAMETRIC MODELING

When creating a parametric model, we use a procedure called *constraining*, which adds flexibility and variability to the designing process while placing some limits on the basic shape. For example, if you need to make adjustments to improve the design and change the model, the constraints allow you to do this without changing the critical shape or required dimensions. Note the following model:

The constraints on this part:
d1 = 3
d0 = d1 (whatever its size)
d2 = d1/2
d3 = d1 x 2

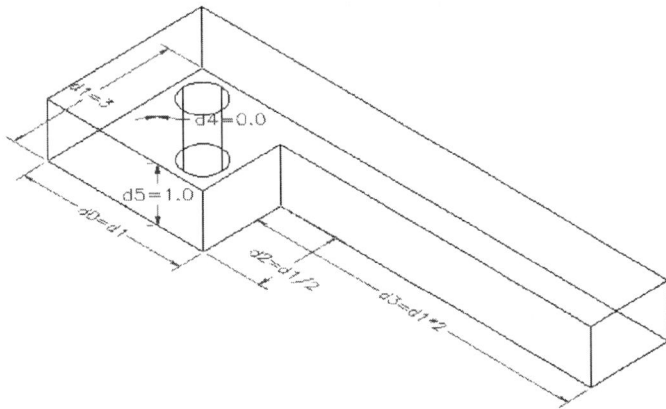

Whatever the size of d1, the other ratios must stay the same.

Note the same basic shape when we change d1 from 3 to 5:

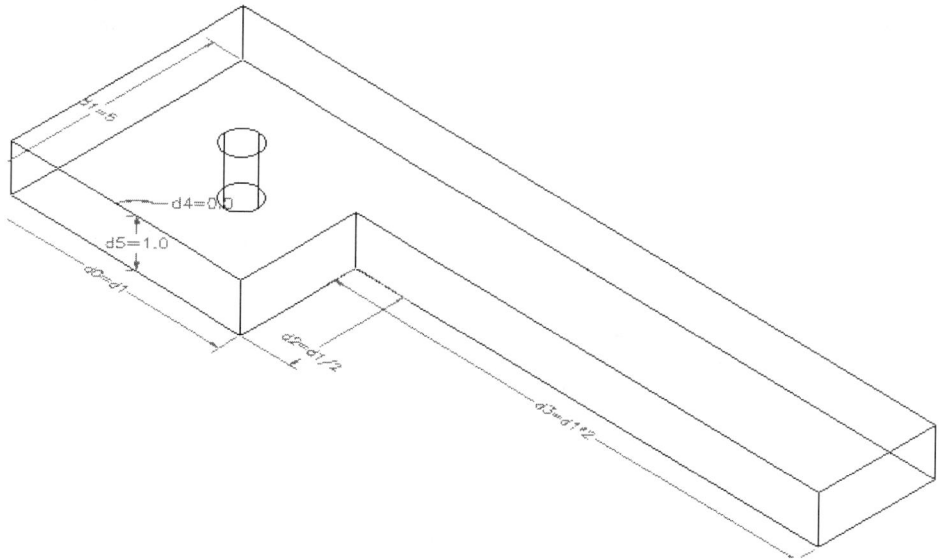

By constraining this way, you can hold some dimensions and ratios the same while experimenting with the overall size of the part.

If you wanted to hold the hole in the same center position, you would constrain the hole as follows:

Constraining the Hole
D1 is the left side of the model
D6 = d1/2
D7 = d1/2

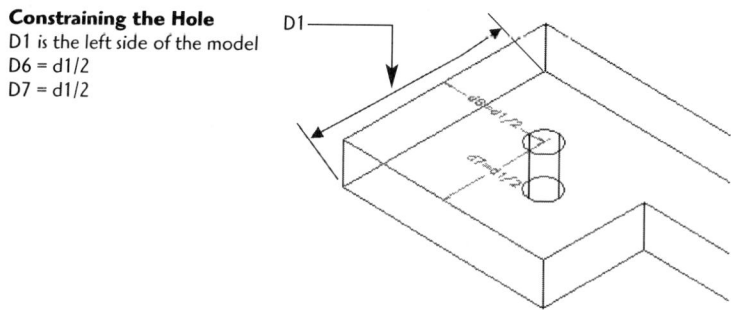

Whatever the size of d1, d6 and d7 will always be half of d1.

THE PRIMARY STEPS IN CONSTRAINING

Fully constraining with Mechanical Desktop requires two steps, **profiling** and **dimensioning**. The *profiling step* examines the shape to assure it is a "closed profile" and capable of becoming a 3D feature. Profiling also cleans up any "roughly drawn" profiles.

The *dimensioning step* adds dimension variables that can be changed and modified as the part design is refined.

CONSTRAINING | 15

Roughly Drawn Shape
Even though a roughly drawn shape, it still must be a closed profile.

Profiled Shape
The hidden lines show the remaining unconstrained sides.

Dimensionally Constrained
As dimensional constraints are added the sides with hidden lines are replaced with solid lines.

Once these two steps are completed, the profile (2D shape) can be further adjusted as needed with the AMMODDIM (modify dimension) command. The dimension is selected, and the original dimension is displayed in the command line.

Command: `Enter Dimension Value <4.375>` *(Enter the new dimension)* ↵.

The sketch updates to the new size.

Now, the model is ready to be turned into a 3D feature. Extruding, revolving, sweeping, and lofting are all methods of turning a profile into a 3D feature.

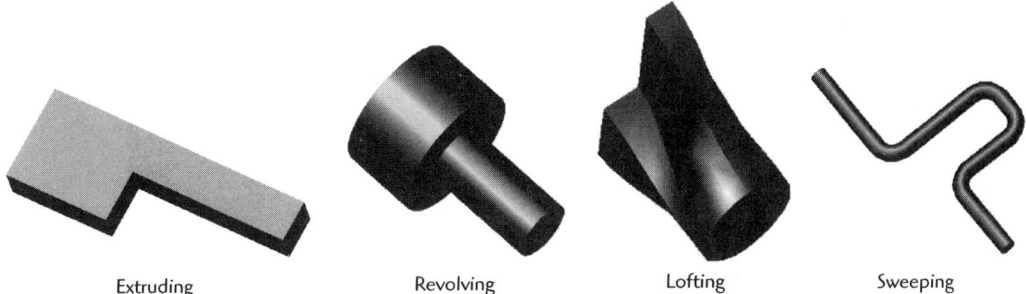

Extruding   Revolving   Lofting   Sweeping

Constraining Project 1

Using standard AutoCAD drawing commands such as PLINE, create a closed profile.

Step 1 Draw a basic shape (profile) and close the profile.

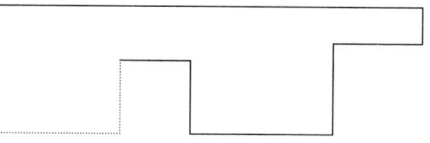

Type **C** ↵ to close the profile.

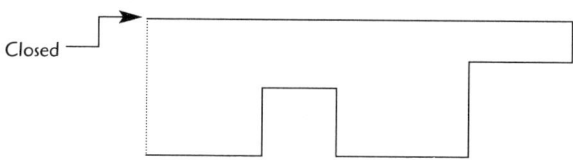

Step 2 Type **AMPROFILE** ↵
or click the Profile icon.
Select the profile ↵.

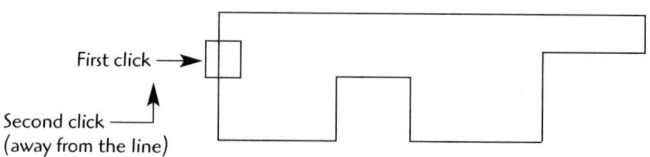

Step 3 Type **AMPARDIM** ↵
or click the Constrain Dimension icon.
Click on a side, then away from the side.

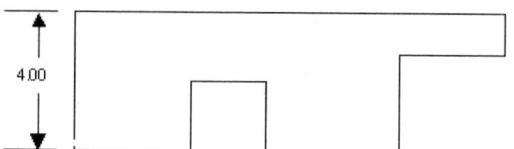

Add the constraining dimension.

CONSTRAINING | 17

Continue to add constraining dimensions until the part is completely constrained. (The message "Solved fully constrained sketch" will appear on the comand line when it is fully contrained.)

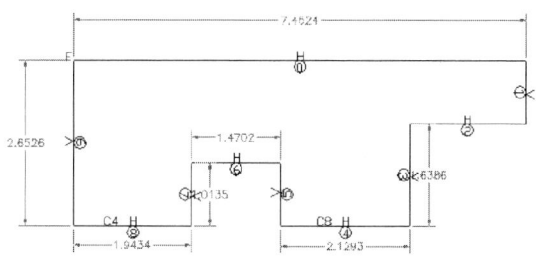

Solved Fully Constrained Sketch

What if I need to **change** one of my constraining dimensions?

Step 4 Type AMMODDIM ↵
or click the Constrain Dimension icon.
— Select the dimension.

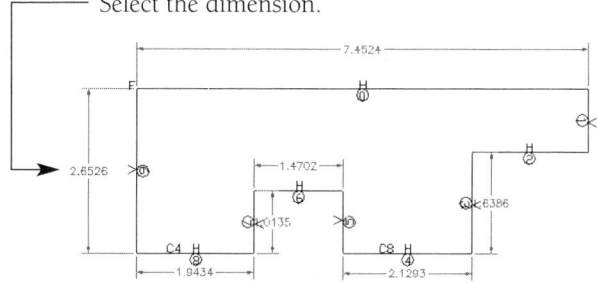

Key in the new constraining dimension ↵.
— The sketch updates to the new dimensions.

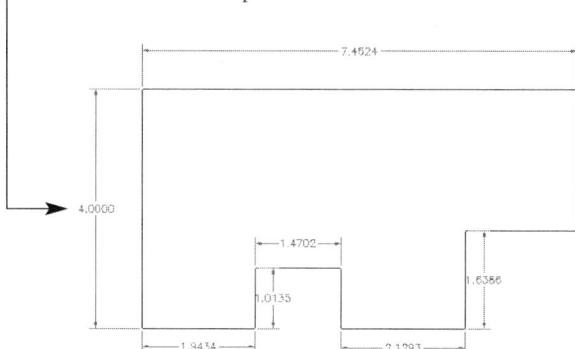

CHAPTER 2

Step 5 Let's get an isometric angle view of the profile.
Click **8** ↵.

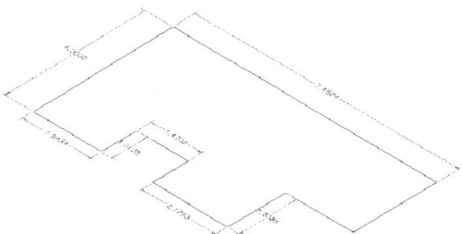

Now let's turn the sketch into a feature (3D feature). For this project you will extrude the sketch into a feature.

Step 6 Type **AMEXTRUDE** ↵
or click the Extrude icon.
Select the sketch profile.

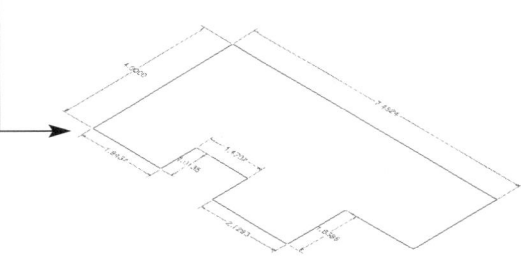

Step 7 Key in the following:

Distance: **1.000**

Draft Angle: **0**

Type: **Blind**

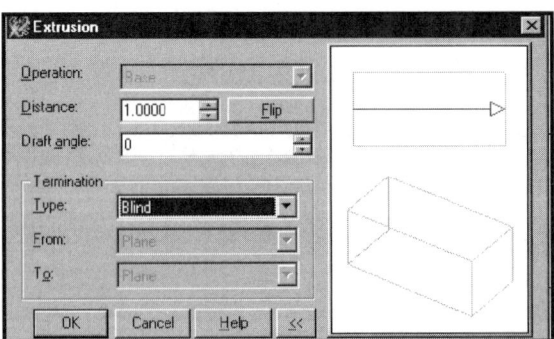

CONSTRAINING | 19

The extruded part can be viewed in several different ways. Clicking on the **Toggle Shading/Wireframe** icon will switch between the wireframe and shaded views.

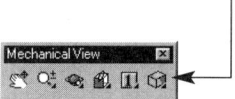

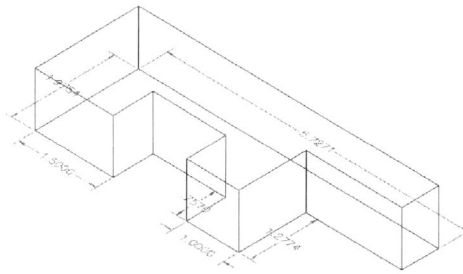
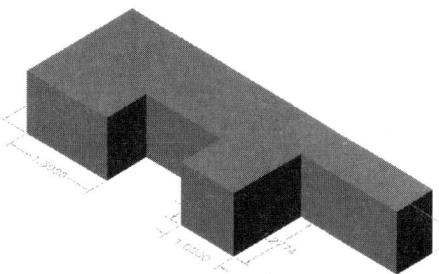

Constraining Project 2

Let's now practice constraining using some basic formulas to keep the sketch in proportion.

Step 1 Roughly sketch this T shape.
(Use Close on the last line.)

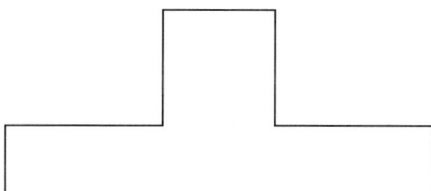

Step 2 Type **AMPROFILE** ↵
or click the Profile icon.
Select the profile.

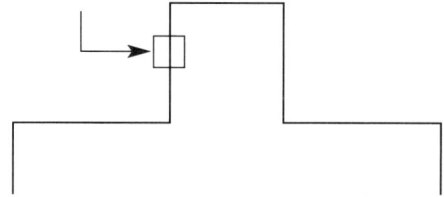

CHAPTER 2

Step 3 Type **AMPARDIM** ↵
or click the Constrain Dimension icon.
Select on the left side of the line, then click away to place the dimension. (Make the dimension 5.0.)

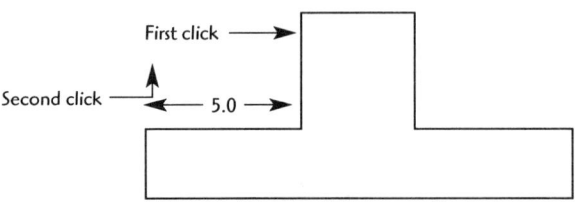

Step 4 Now you will view the dimensions as equations. (Don't let the term *equations* worry you, you do not have to be a math whiz for this procedure!)

Part
➢ Dimensioning, ➢ Dimension As Equations

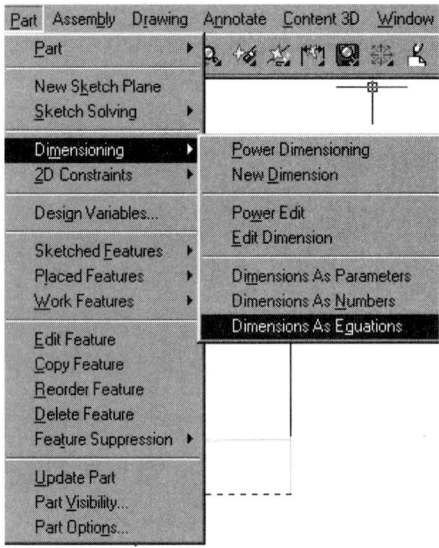

Step 5 Type **AMPARDIM** ↵
or click the Constrain Dimension icon.
Pick the other side of the T to constrain.

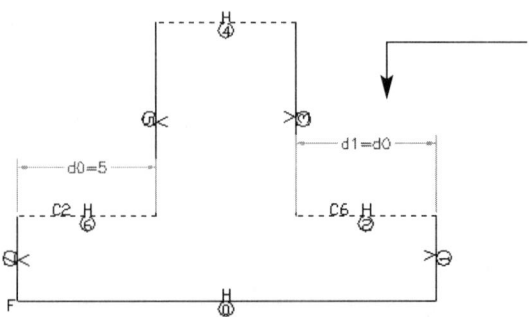

Type =d0 ↵ (indicating this d1 dimension will always equal d0).
Note: Both sides are now 5.0.

Step 6 Continue by selecting the base (bottom) line.
Type =d0*3 ↵.
(*Note:* * is the multiplication symbol.)

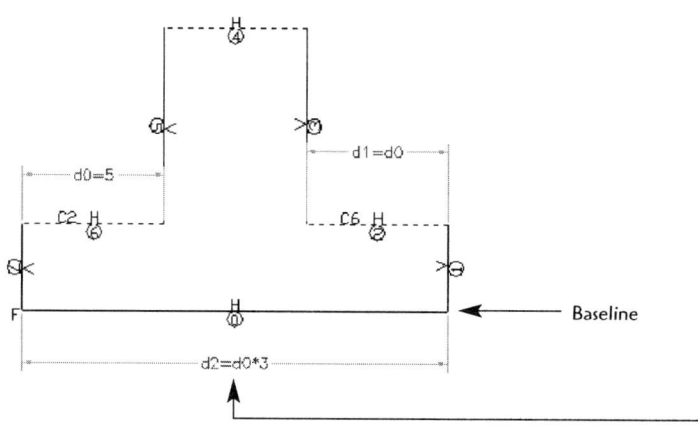

Step 7 Select the vertical side.
Type in =d0/2 ↵.
(*Note:* the / "back slash" is the division symbol.)

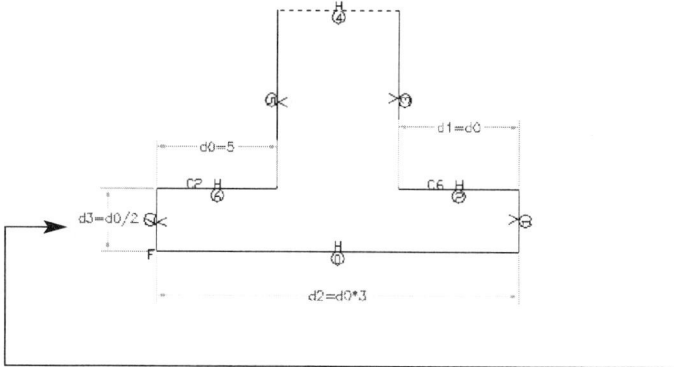

Step 8 Select the vertical side.
Type =d0*2 ↵.
The sketch is now fully constrained with equations.

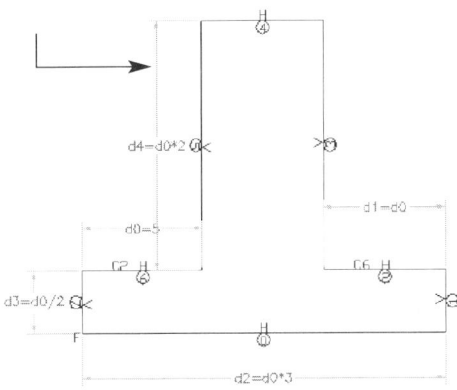

Step 9 You will now edit the **d0** constraint and see how the profile reacts.
Type `AMMODDIM` ↵
or click the Constrain Dimension icon.
Click on the d0=5 dimensional constraint and type in 7 ↵. (*Note:* The dimension changes, but the overall shape is constrained.)

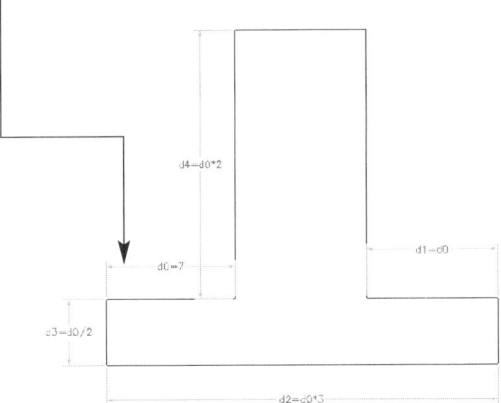

For a Deeper Understanding about
Constraining

Remember there are two primary steps in the constraining process, profiling and dimensional constraints. Profiling examines and adjusts the shape if necessary. Dimensional constraining applies dimensional constraints or formulas to the profile. You may have also noticed the **H** (horizontal) and **V** (vertical) markings as well as the circled numbers added to the profiled lines. These indicate the type of constraint applied to the profile constraints. For example, the **F** indicates a fixed point that cannot move as other dimensions are changed or adjusted. An **H** will hold the line to the horizontal axis, and a **V** will hold a line to the vertical axis. The numbers 0, 1, 2, 3,... are the variable numbers assigned to each element. These numbers can be used if equations are applied to the element.

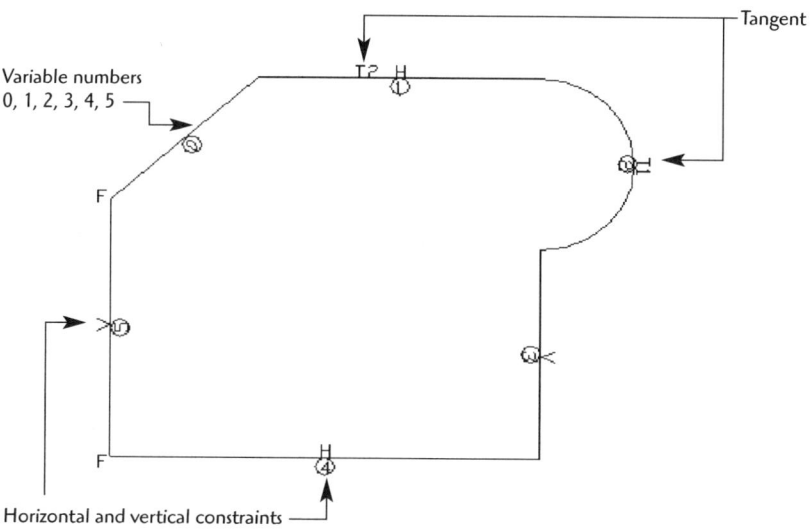

You can make adjustments to your profile by deleting constraints with the **AMDELCON** command or by clicking the **Delete Constraint** icon.

To add constraints use the **AMADDCON** command. This process is not always necessary but can be used for special profile applications and shapes.

To display the existing constraints on a profile use the **AMSHOWCON** command or click the **Show Constraint** icon.

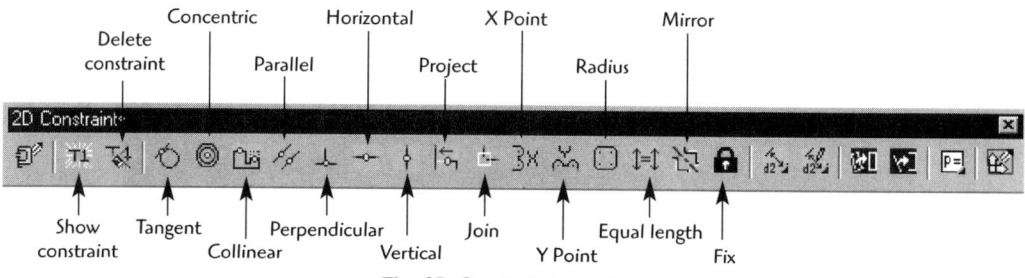

The 2D Constraints toolbar

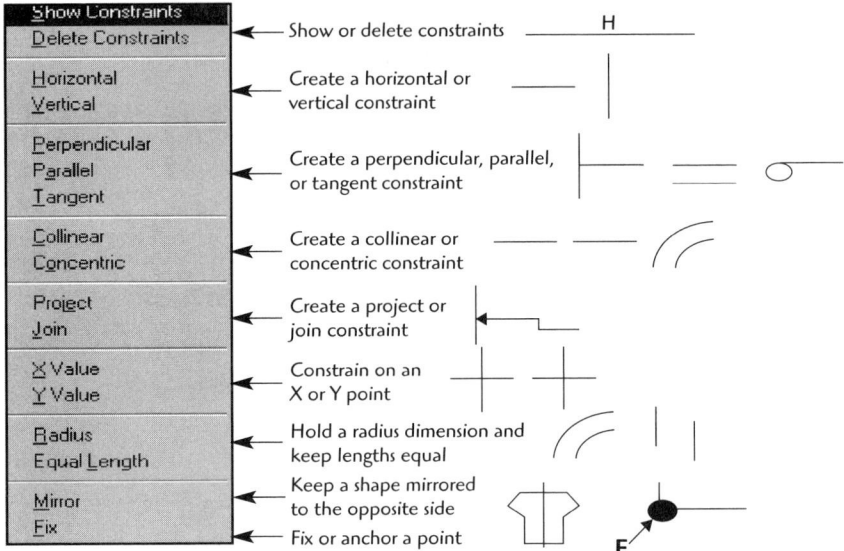

Constraints from the pull-down menu

Drawing Project 2-1

T-Block

Model a T-block to the exact dimensions indicated on the first T-block drawing. Fully constrain it so dimensions can be adjusted. Save the first drawing under a unique name. Now, adjust the constraint dimensions to match the second T-block drawing dimensional sizes. Save the second drawing under a unique name. You should be able to produce the second T-block by changing only one or two constraints. When you have finished, click **8** for an isometric view, and print out both copies for your instructor.

First T-block

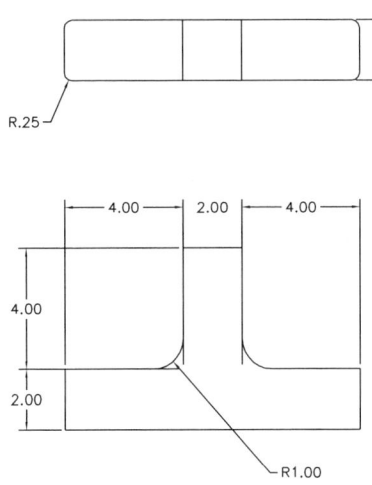
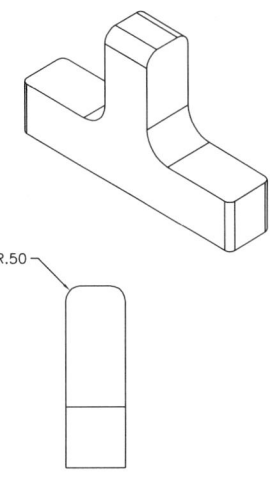

Second (adjusted) T-block

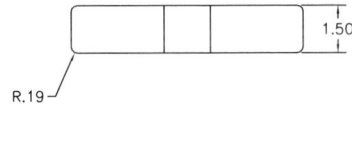
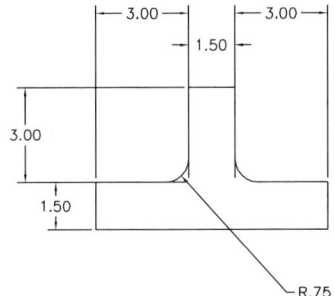
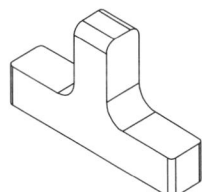
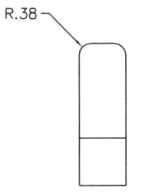

Drawing Project 2–2

Y-Block

Model a Y-block to the exact dimensions indicated on the first Y-block drawing. Fully constrain it so dimensions can be adjusted. Save the first drawing under a unique name. Now, adjust the constraint dimensions to match the second Y-block drawing dimensional sizes. Save the second drawing under a unique name. You should be able to produce the second Y-block by changing only one or two constraints. When you have finished, click the **8** for an isometric view, and print out both copies for your instructor.

First Y-block

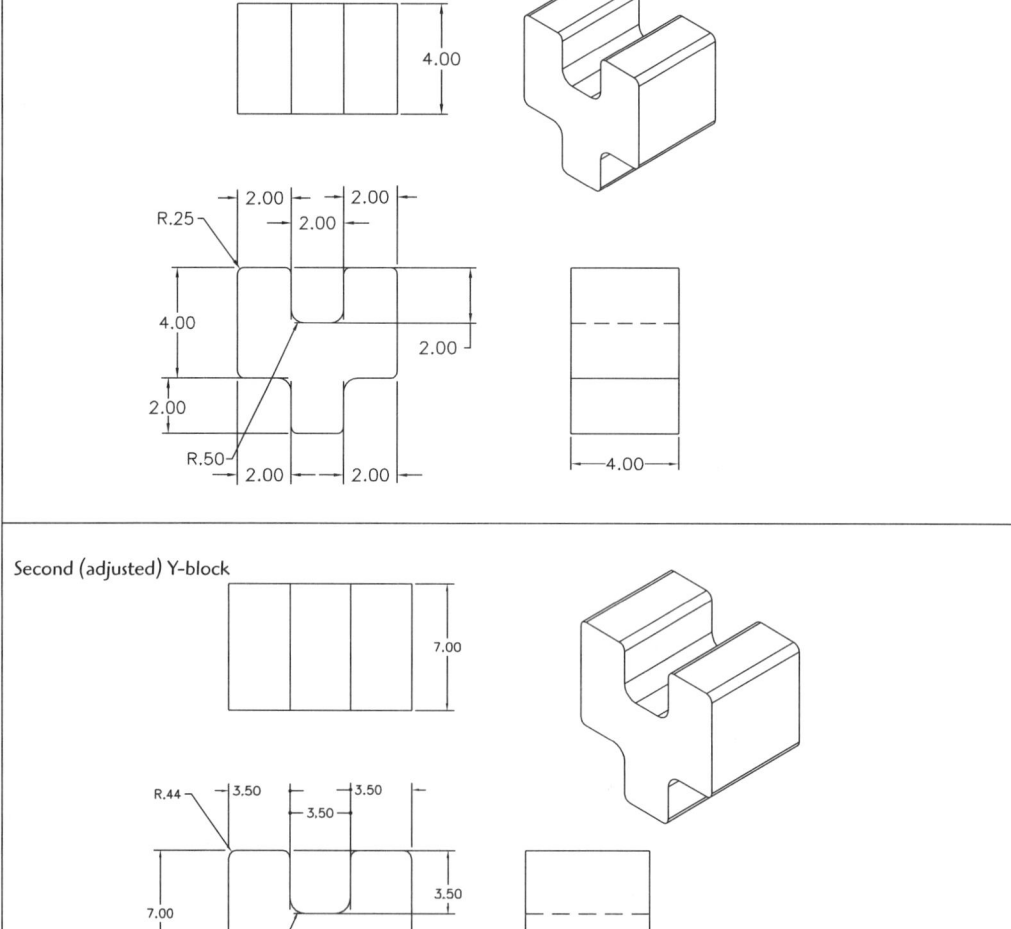

Second (adjusted) Y-block

3D MODELS: TURNING A CONSTRAINED SKETCH INTO A 3D FEATURE MODEL

Once a sketch is fully constrained, there are a variety of commands for turning the sketch into a 3D feature model. **Extrude**, **Revolve**, **Sweep**, and **Loft** are typical commands. The Extrude command is the simplest and most obvious of the 3D commands. To create an extrusion, the 3D thickness is extruded a distance off the sketched profile. A profile may be extruded in either direction or in both directions (known as a midplane extrusion).

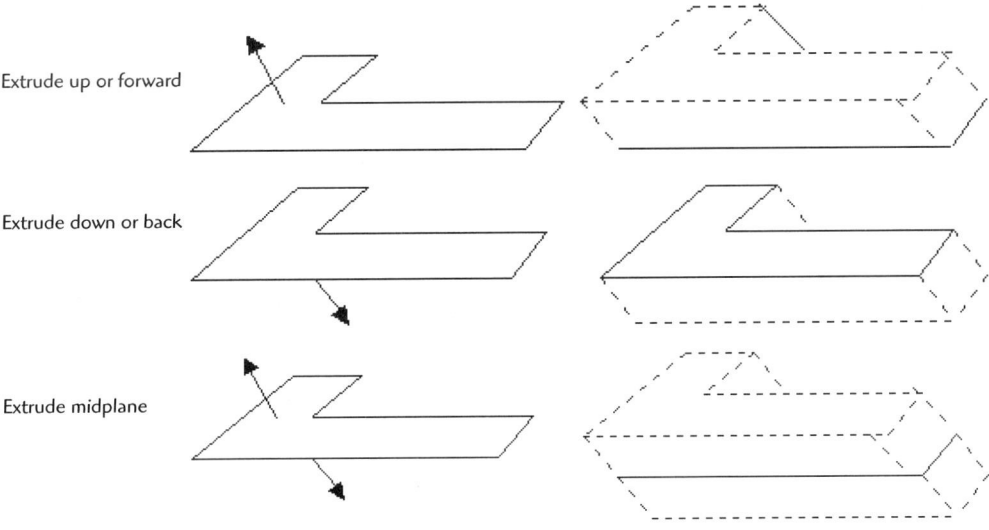

Extrude Project 1

Step 1 Sketch a profile, remembering to close it.

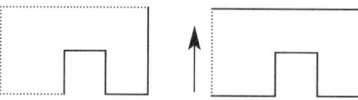

Step 2 Type **AMPROFILE** ↵
or click the Profile icon.
Select profile ↵.

Step 3 Type **AMPARDIM** ↵
or click the Constrain Dimension icon.
(Constrain the sides until the profile is fully constrained.)

Step 4 Type 8 ↵ to generate an isometric view.

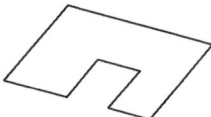

Step 5 Type **AMEXTRUDE** ↵
or click the Extrude icon.

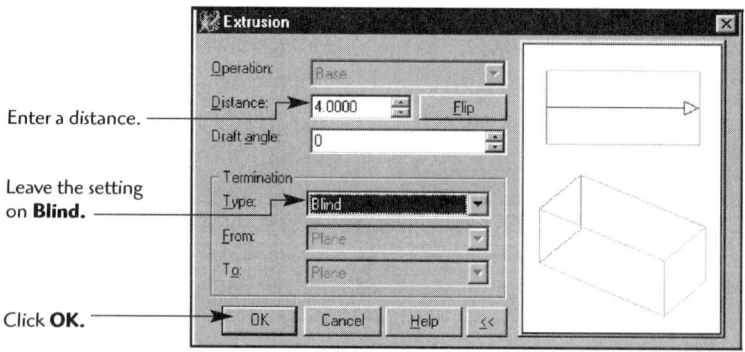

Enter a distance.

Leave the setting on **Blind.**

Click **OK.**

3D MODELS | 29

You now see the extruded 3D Feature.

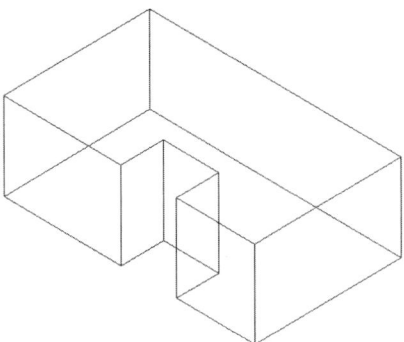

Step 6 Let's now examine the model. (Click the Render/Wireframe icon and dynamically rotate the model.)

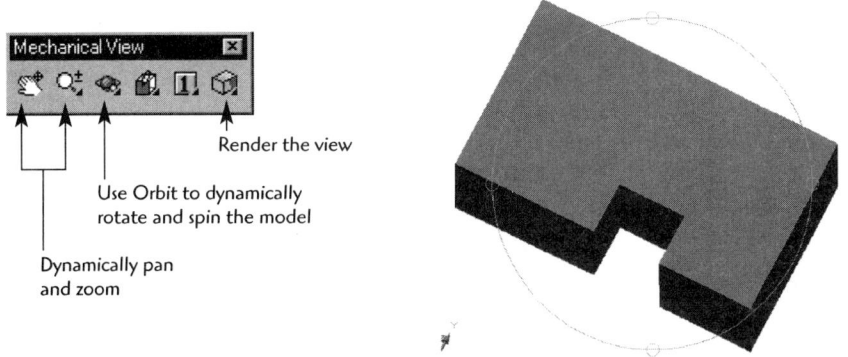

For a Deeper Understanding about
Extruding

The Extrusion dialog box will have different options depending on whether it is the first profile being extruded (this is called a Base part), or if the sketch is an add-on to the original base part. Note the additional Operation options available.

The operation **Cut** will cut or remove material from the base part. Make sure the arrow points into the base material, not away.

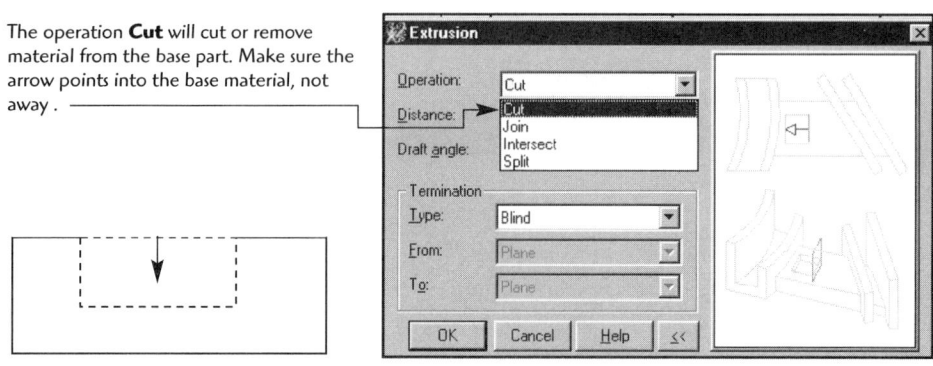

An extrusion set to cut into the base material

The **Join** operation "welds" the shape onto the base part.

The **Distance** sets how far the joined extrusion will extend.

Enter a **draft angle** *only* if you want a tapered extrusion.

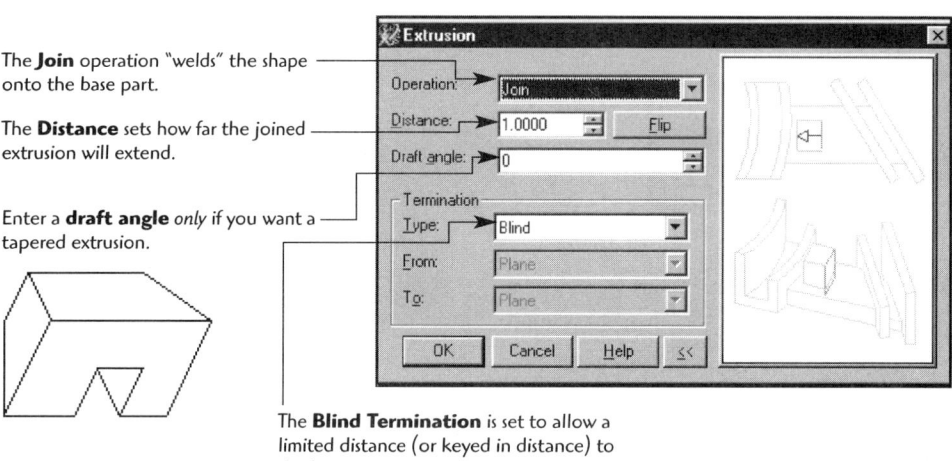

The **Blind Termination** is set to allow a limited distance (or keyed in distance) to the Join operation.

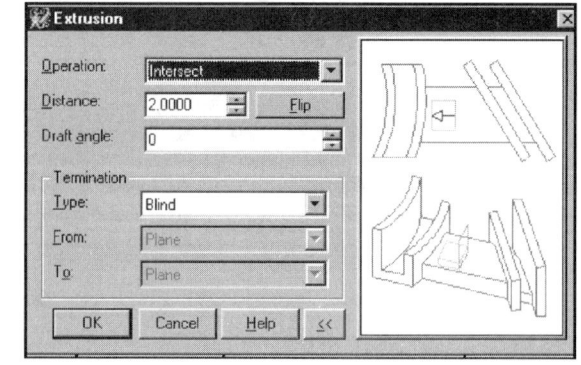

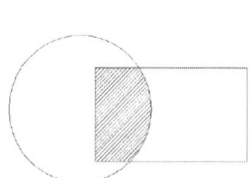

The **Intersect** operation produces a shape that encloses only the area shared by both profiles.

3D MODELS | 31

A round and a rectangular shape that partially overlap

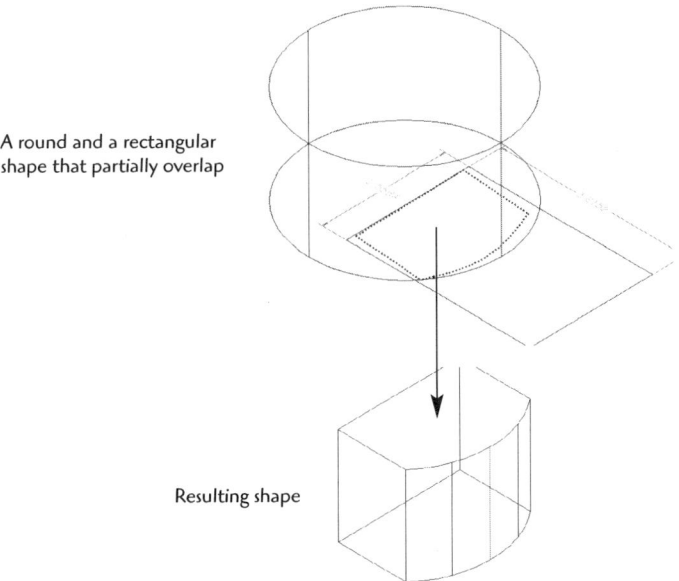

Resulting shape

The **Split** option allows you to cut a new and separate part using a profile.

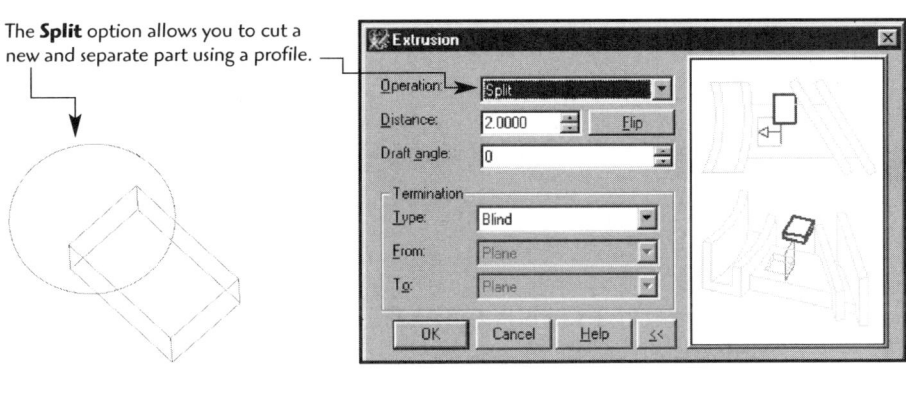

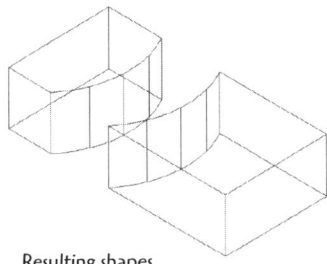

Resulting shapes

The **Termination** options define where the extrusion will terminate.

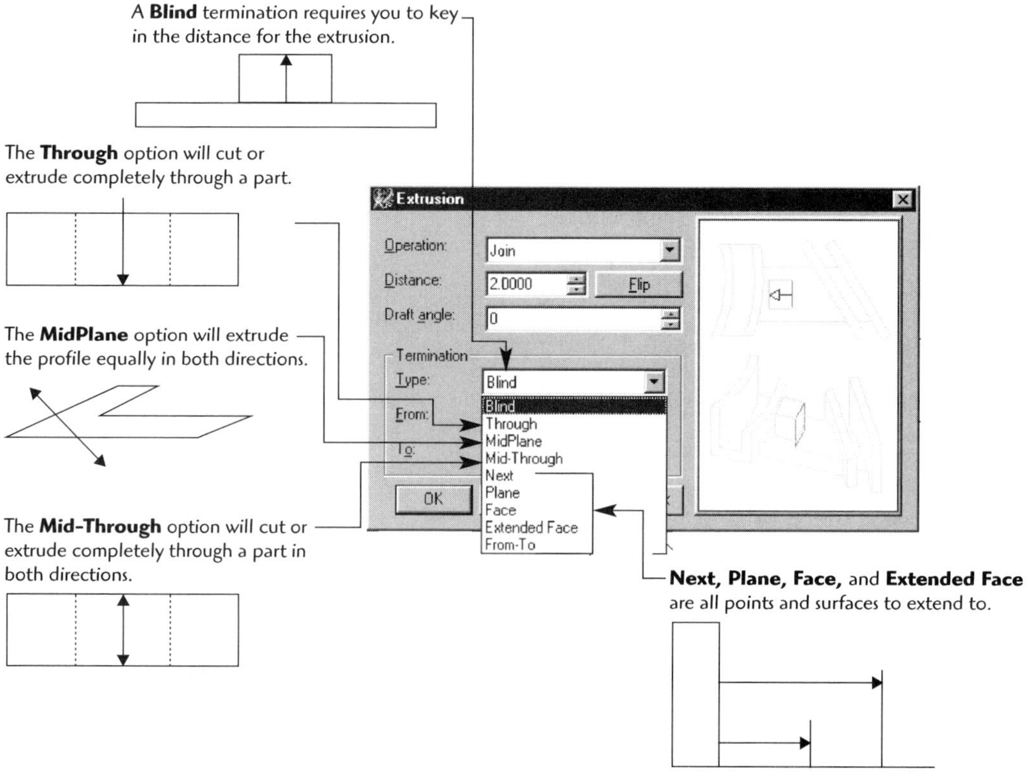

A **Blind** termination requires you to key in the distance for the extrusion.
The **Through** option will cut or extrude completely through a part.
The **MidPlane** option will extrude the profile equally in both directions.
The **Mid-Through** option will cut or extrude completely through a part in both directions.
Extrusion
Operation:
Join
Distance:
2.0000
Flip
Draft angle:
0
Termination
Type:
Blind
From:
To:
OK
Blind
Through
MidPlane
Mid-Through
Next
Plane
Face
Extended Face
From-To
Next, Plane, Face, and **Extended Face** are all points and surfaces to extend to.

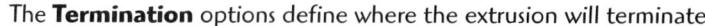

ADDING HOLES, FILLETS, AND CHAMFERS TO A 3D FEATURE MODEL

Adding Holes

Adding holes is a straightforward process. The procedure involves defining the hole sizes, then defining the location. Three typical hole operations are possible:

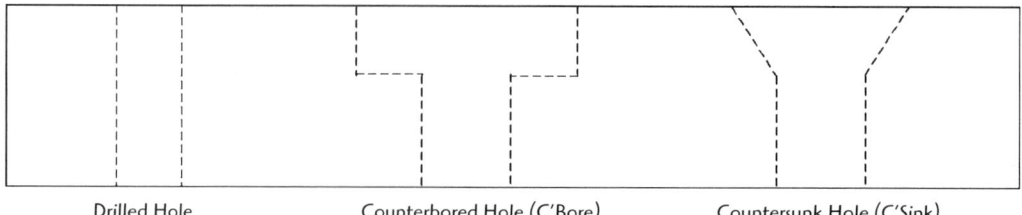

Drilled Hole   Counterbored Hole (C'Bore)   Countersunk Hole (C'Sink)

Counterbored holes are usually used for round head screws or fasteners. **Counter-sunk** holes are usually used for flat head screws or fasteners.

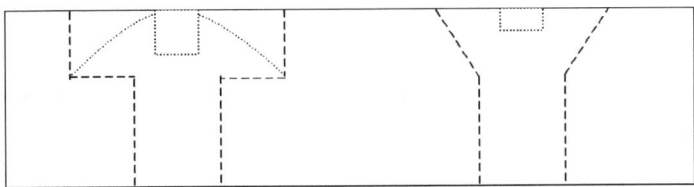

Hole Project

Step 1   Sketch, fully constrain, and extrude a shape, or open a previously created part.

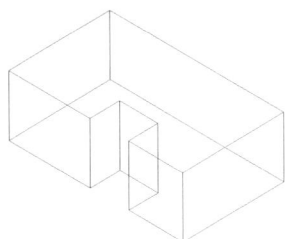

Step 2   Type **AMHOLE** ↵.
Set the values in the Hole dialog box, and click **OK**. You may use the default settings.

Use **Tap** only if you want to add threads to the hole.

The operation can be a drilled, counterbored, or countersunk hole.

The **termination** can be set to go **Through** the part, or set to **Blind**, which means a specified depth. You can also set the depth to stop at a plane or surface.

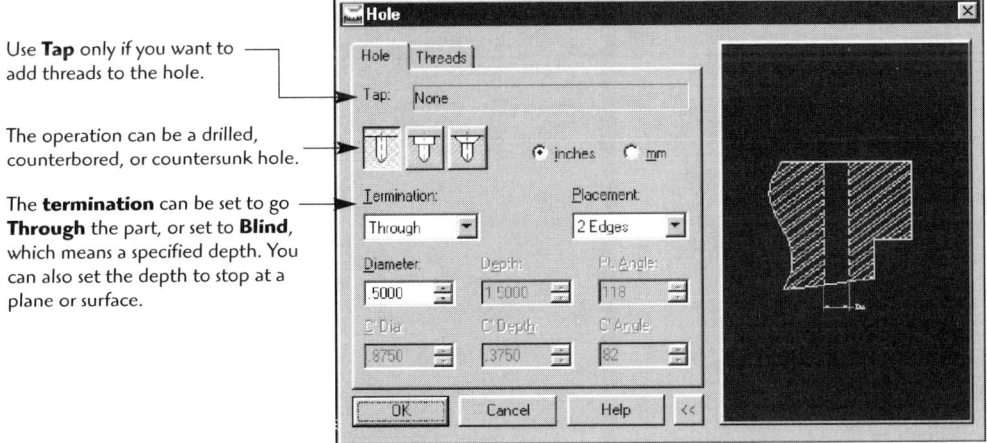

The **placement** of the hole can be defined by **2 Edges** or a **Concentric** radius or circle. Additionally, points or other holes can define a point for a hole.

Diameter dimensions as well as counterbores and countersink dimensions are added as needed.

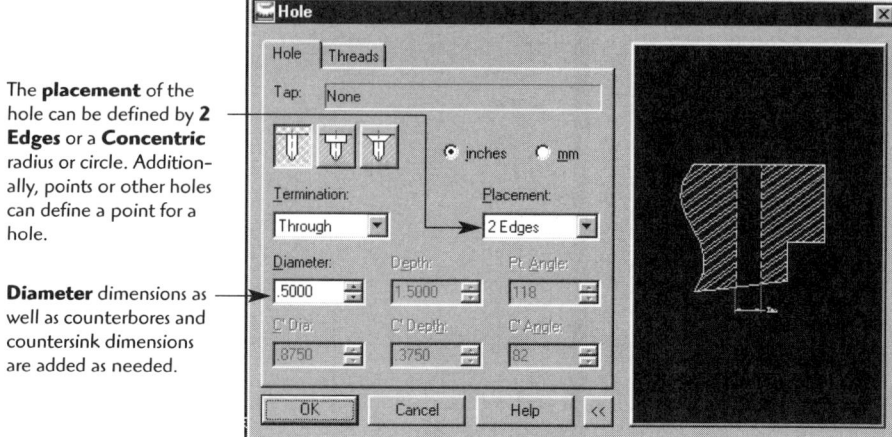

Step 3 With the **2 Edges** option set in the Placement box, select two edges on the model, then drag them to the placement point.

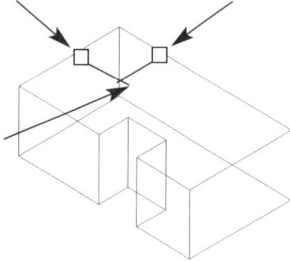

The Command line will ask you to confirm each distance from the two edges.

Enter Distance from first edge *Type in the distance* ↵.

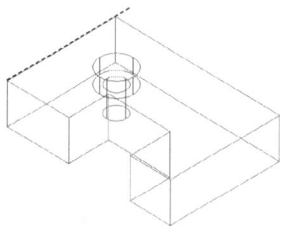

`Enter Distance from second edge` *Type in the distance* ↵.

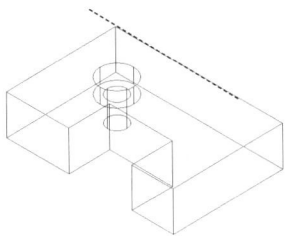

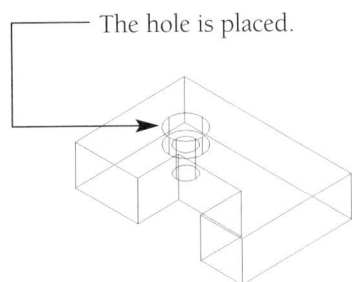

The hole is placed.

Adding Fillets

Fillets are commonly added to strengthen a part by reducing the "notch effect" and resulting stress riser. The fillet may also improve the part cosmetically.

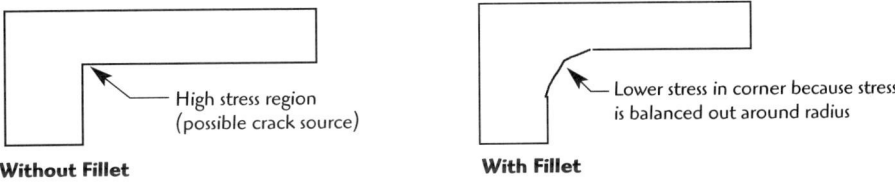

High stress region (possible crack source)

Without Fillet

Lower stress in corner because stress is balanced out around radius

With Fillet

Fillet Project

Step 1 Sketch, fully constrain, and extrude a shape or open a previously created part.

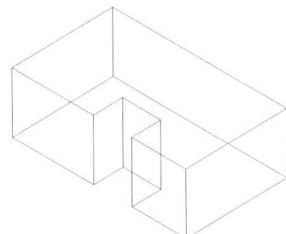

36 | CHAPTER 3

Step 2   Type **AMFILLET** ⏎.
   Set the values in the Fillet dialog box, and click **OK**. You may use the default settings.

For standard fillets, leave the Constant option selected and enter the Radius value.

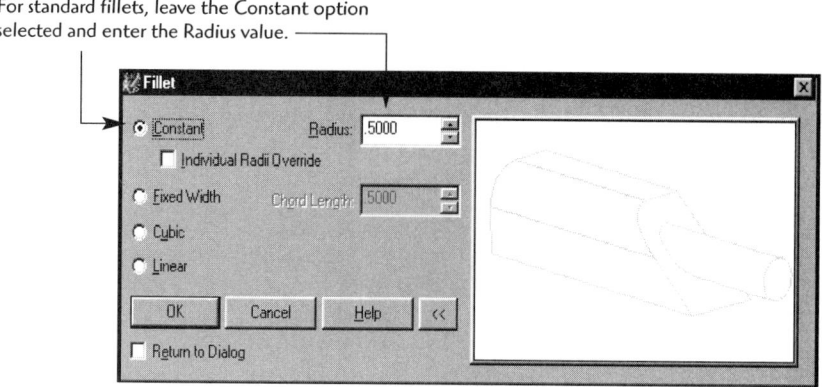

Step 3   Select the edge to be filleted ⏎.

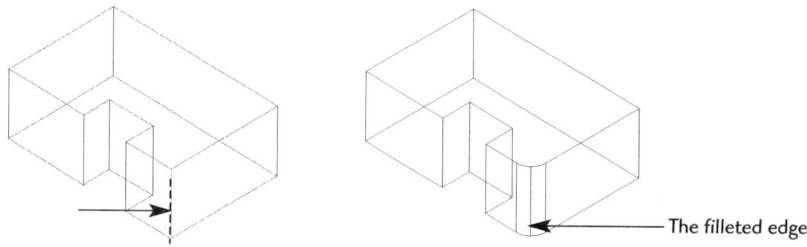

The filleted edge

Adding Chamfers

Step 1   Sketch, fully constrain, and extrude a shape or open a previously created part.

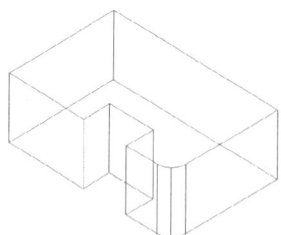

3D MODELS | 37

Step 2   Type **AMCHAMFER** ↵.
Set the values in the Chamfer dialog box, and click **OK**. You may use the default settings.

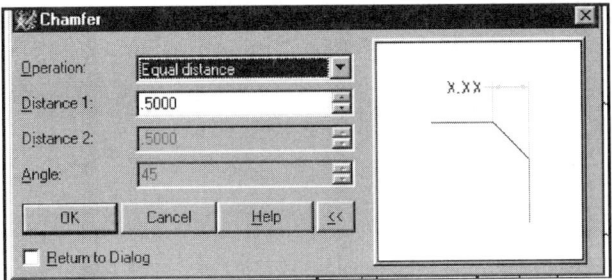

Step 3   Click the edge to have the chamfer applied ↵.

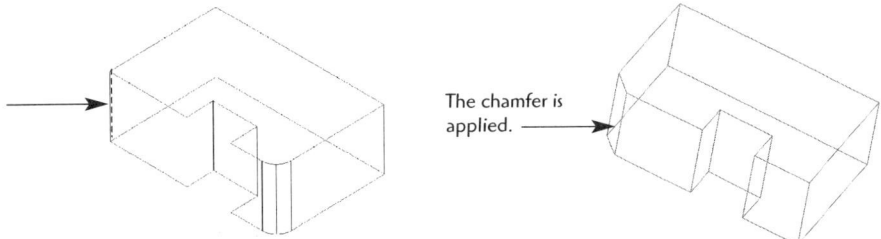

The chamfer is applied.

For a Deeper Understanding about
Fillets and Chamfers

Additional fillet options include:

- **Fixed Width**
- **Cubic**
- **Linear**

The *Fixed Width* option has a *chord length* instead of a radius due to the variable shape of a radius on this type of fillet.

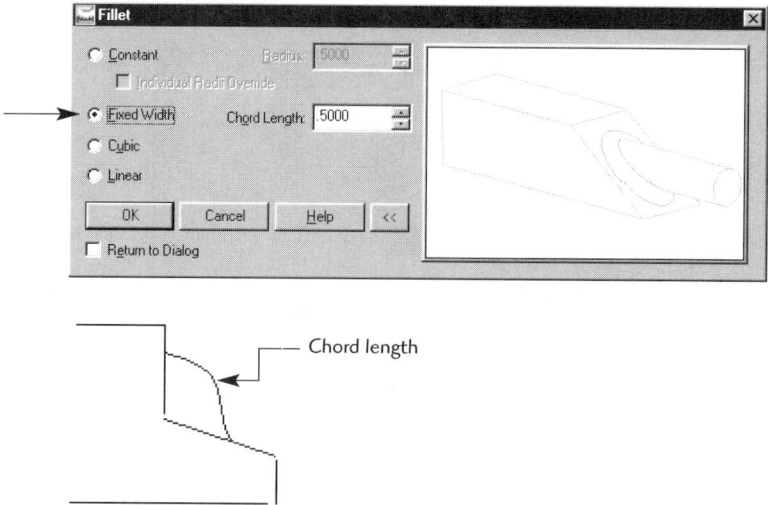

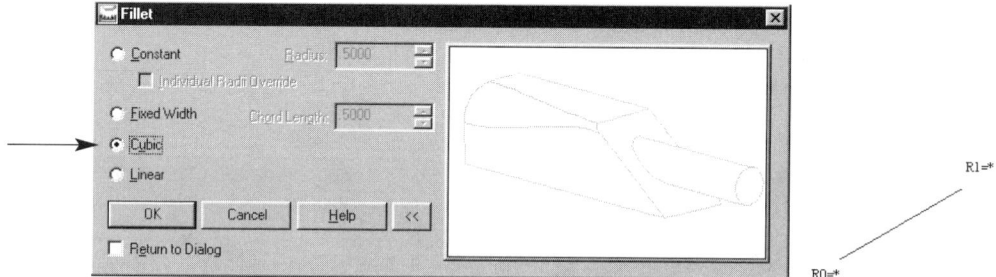

Chord length

The *Cubic* option produces a gradual spline type fillet. A radius is entered for both ends. The result is a smooth transition fillet.

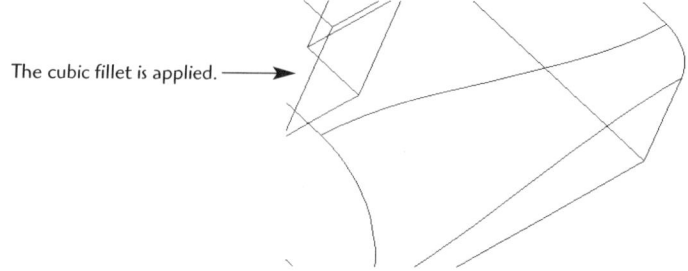

After selecting this option, click on each R and enter a radius value, then ↵.

The cubic fillet is applied.

The *Linear* option produces a fillet with a straight transition between two radius points. A radius is entered for both ends.

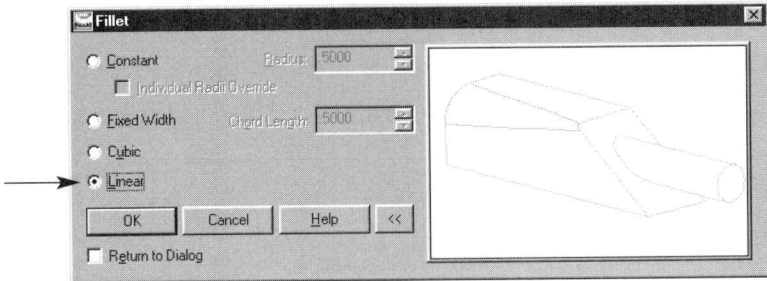

After selecting this option, click on each R and enter a radius value, then ↵.

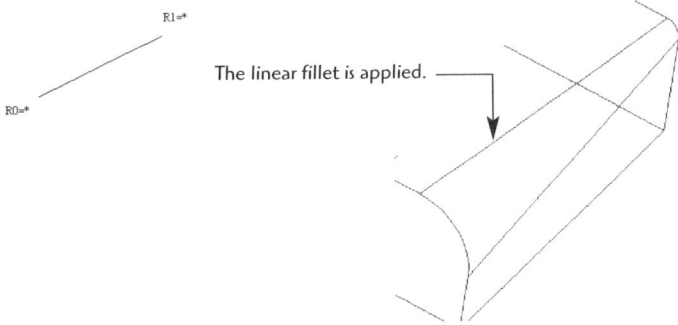

The linear fillet is applied.

Additional options for applying chamfers include:

- **Equal distance**
- **Two distances**
- **Distance and angle**

The *Equal distance* option creates a chamfer with two equal sides. The chamfer is entered in Distance 1:.

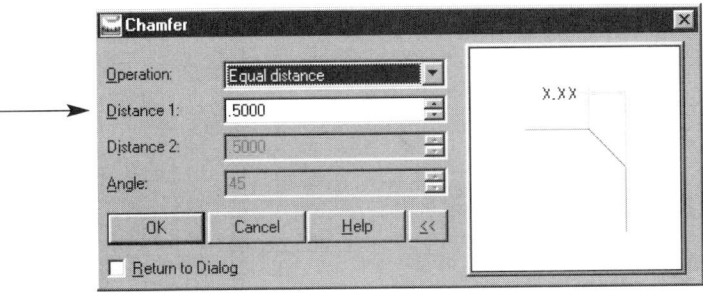

When a chamfer with unequal sides is desired, the *Two distances* option is selected. Distance 1: and Distance 2: must be entered.

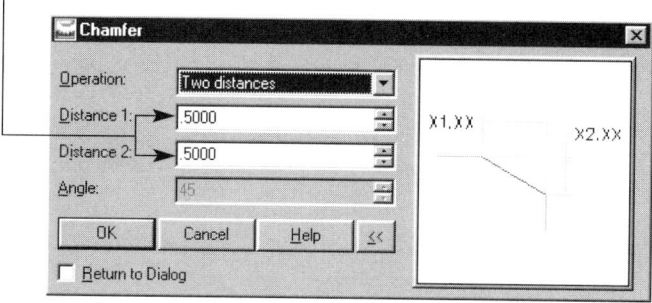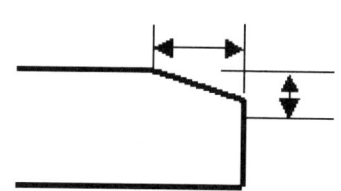

When a distance and desired angle are known, the *Distance and angle* option is selected.

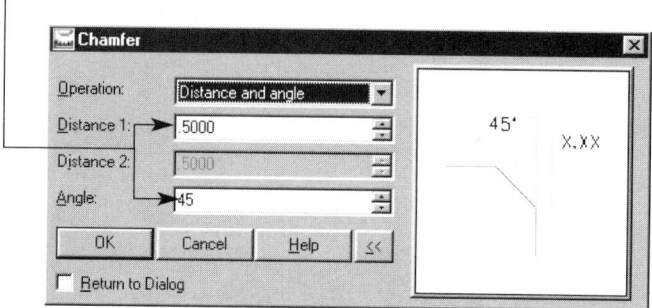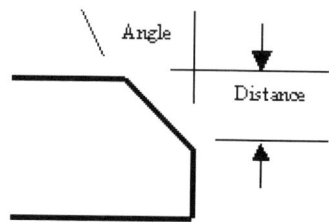

VIEWING 3D MODELS

Although we discussed this topic previously, let's take a detailed look at how to view and examine models. Make sure the Mechanical View toolbar is displayed. If it is not, from the pull-down menus select:

View
> Toolbars
>> Desktop Toolbars
>>> Mechanical View

Or type **Toolbar** ↵, click the **AMDTPP** menu group, and select **Mechanical View**.

3D MODELS | *41*

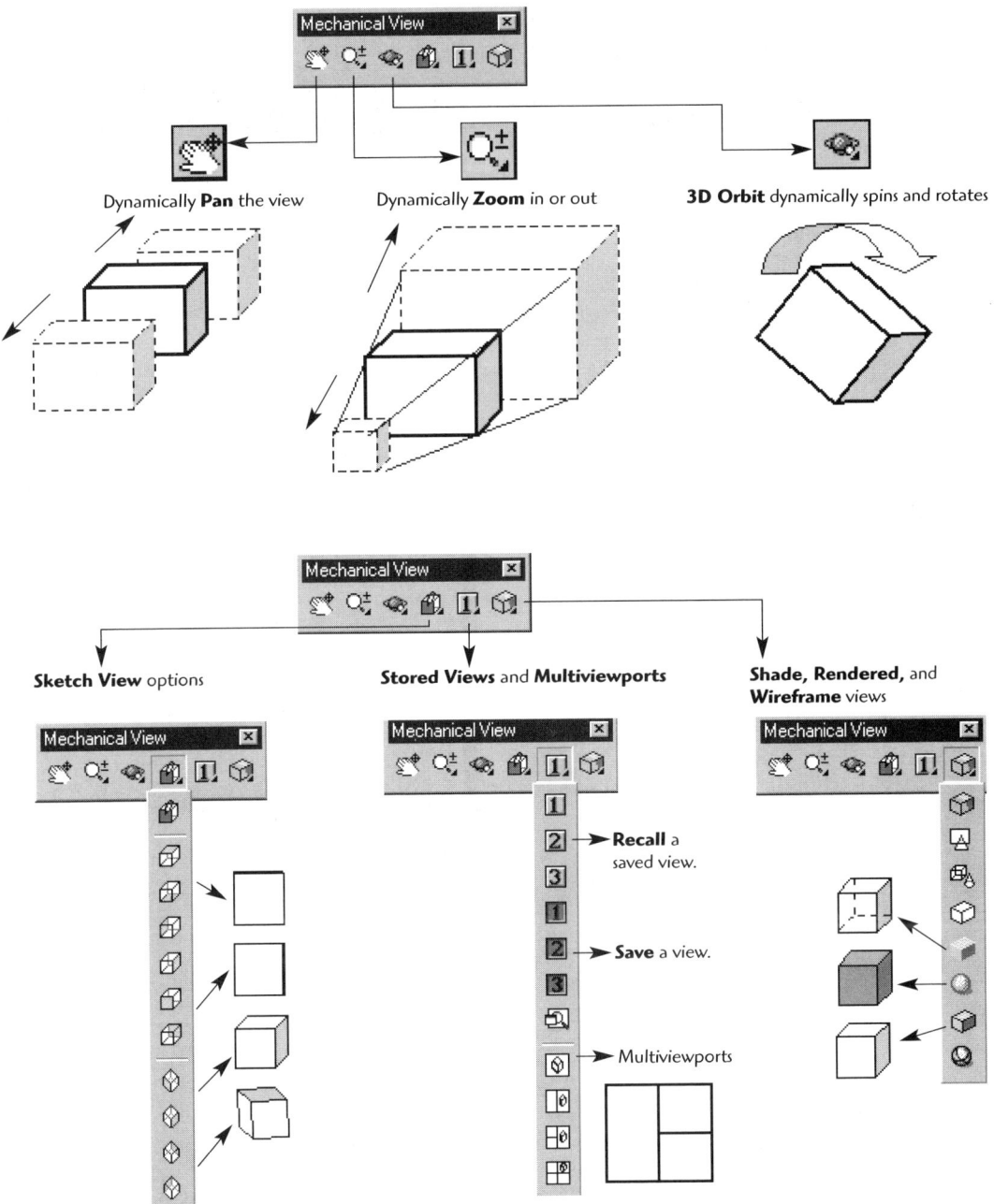

Dynamically **Pan** the view

Dynamically **Zoom** in or out

3D Orbit dynamically spins and rotates

Sketch View options

Stored Views and **Multiviewports**

Shade, Rendered, and **Wireframe** views

Recall a saved view.

Save a view.

Multiviewports

For a Deeper Understanding about
Viewing Models

There are a number of viewing options available with **flyout** buttons.
For example, the zoom options flyout lets you select from a variety of zoom methods:

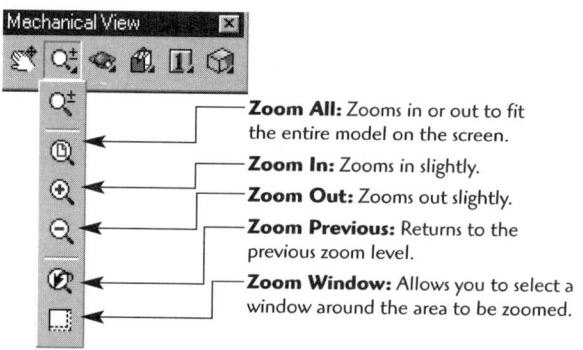

Zoom All: Zooms in or out to fit the entire model on the screen.
Zoom In: Zooms in slightly.
Zoom Out: Zooms out slightly.
Zoom Previous: Returns to the previous zoom level.
Zoom Window: Allows you to select a window around the area to be zoomed.

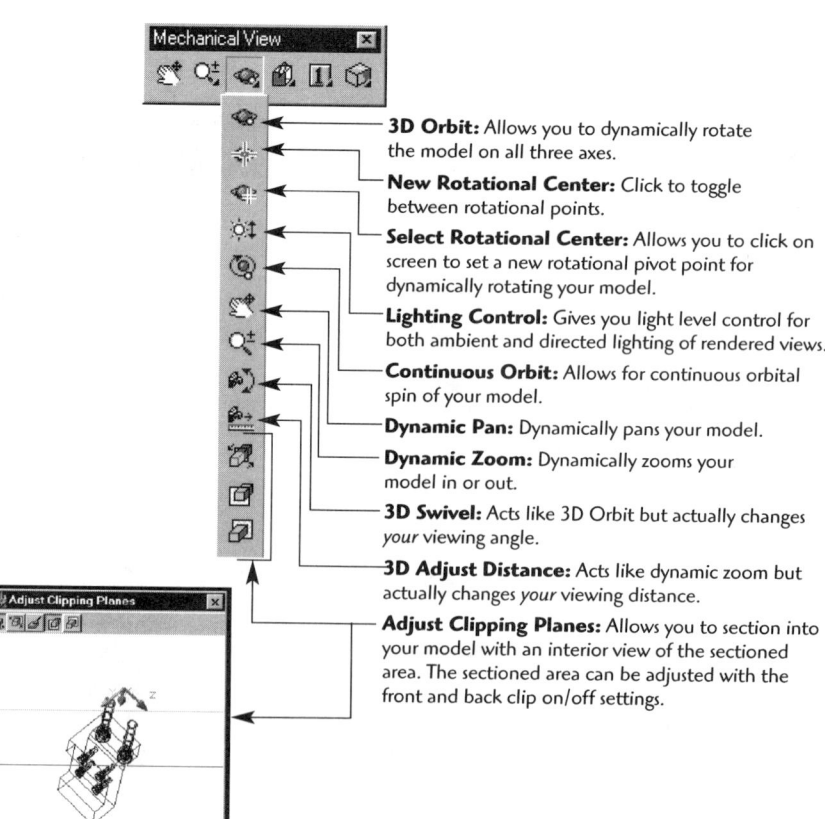

3D Orbit: Allows you to dynamically rotate the model on all three axes.
New Rotational Center: Click to toggle between rotational points.
Select Rotational Center: Allows you to click on screen to set a new rotational pivot point for dynamically rotating your model.
Lighting Control: Gives you light level control for both ambient and directed lighting of rendered views.
Continuous Orbit: Allows for continuous orbital spin of your model.
Dynamic Pan: Dynamically pans your model.
Dynamic Zoom: Dynamically zooms your model in or out.
3D Swivel: Acts like 3D Orbit but actually changes *your* viewing angle.
3D Adjust Distance: Acts like dynamic zoom but actually changes *your* viewing distance.
Adjust Clipping Planes: Allows you to section into your model with an interior view of the sectioned area. The sectioned area can be adjusted with the front and back clip on/off settings.

3D MODELS | 43

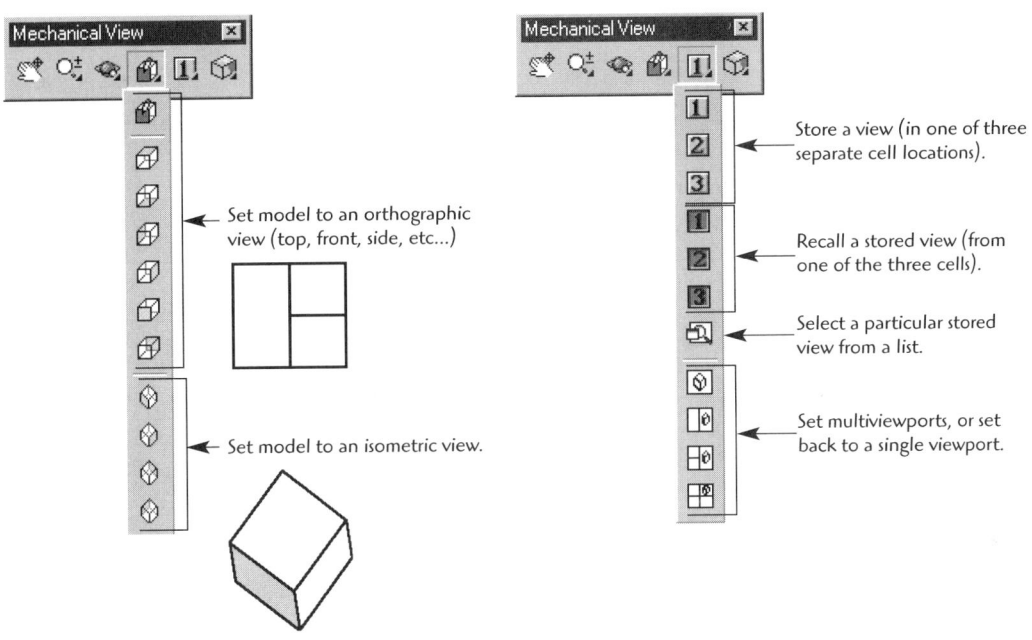

Set model to an orthographic view (top, front, side, etc...)

Set model to an isometric view.

Store a view (in one of three separate cell locations).

Recall a stored view (from one of the three cells).

Select a particular stored view from a list.

Set multiviewports, or set back to a single viewport.

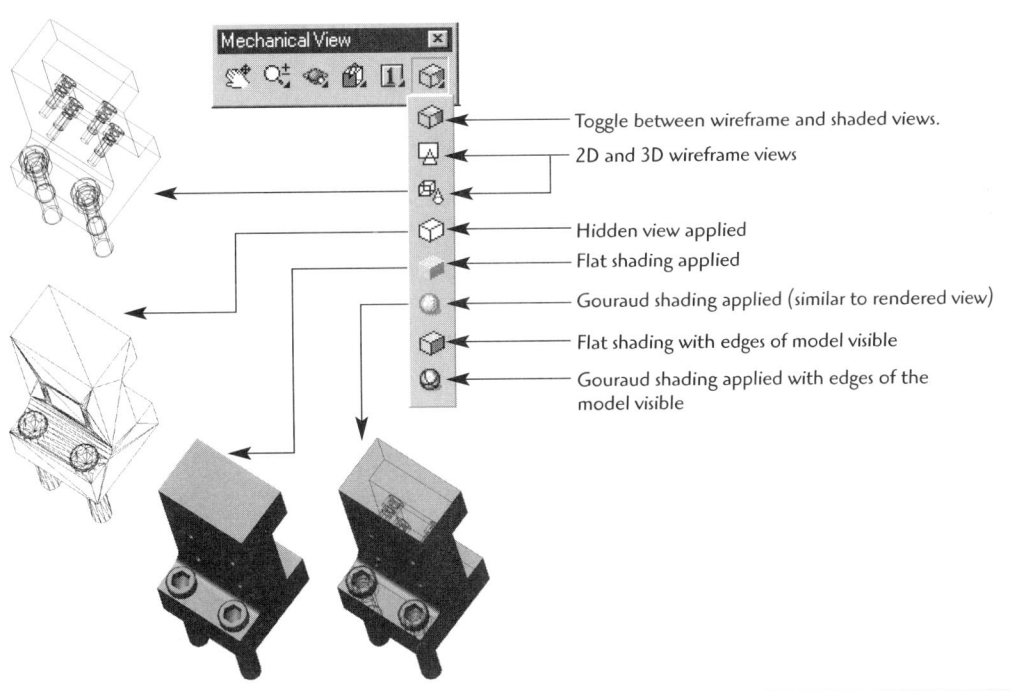

Toggle between wireframe and shaded views.
2D and 3D wireframe views
Hidden view applied
Flat shading applied
Gouraud shading applied (similar to rendered view)
Flat shading with edges of model visible
Gouraud shading applied with edges of the model visible

HANDY SINGLE-KEY VIEWING SHORTCUTS

You may find using a single key to adjust or switch your views is the fast and easy method for navigating. The following helpful keys will help speed up your designing process.

Use the **9** key for the **plan** view.

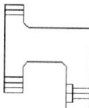

Use the **8** key for the **iso** view.

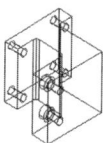

Use the **7** key for the **world** view.

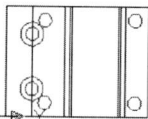

Use the **6** key for the **front** view.

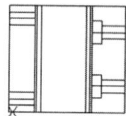

Use the **5** key for the **top** view.

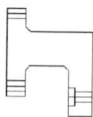

Use the **4** key for the **four viewports** view.

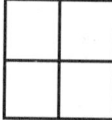

[3]↵   Use the **3** key for the **three viewports** view.

[2]↵   Use the **2** key for the **two viewports** view.

[1]↵   Use the **1** key for the **one viewport** view.

Drawing Project 3-1

Complete T-Block with Holes

Model the T-block to the exact dimensions indicated. Add the fillets and holes as shown. When you have finished, save the drawing and click **8** for an isometric view. Print out a copy for your instructor.

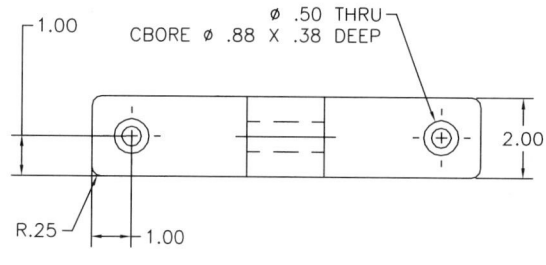
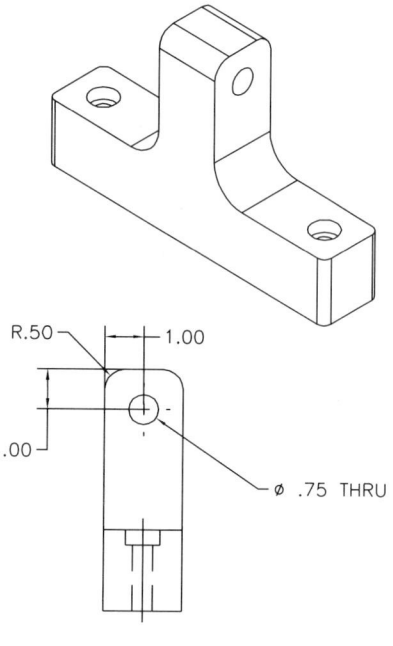
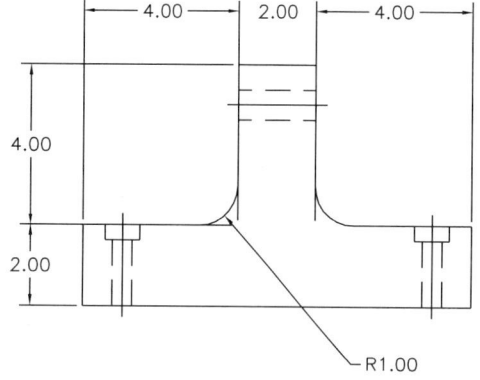

Drawing Project 3-2

Complete Y-Block with Holes

Model the Y-block to the exact dimensions indicated. Add the fillets and holes shown. When you have finished, save the drawing and click **8** for an isometric view. Print out a copy for your instructor.

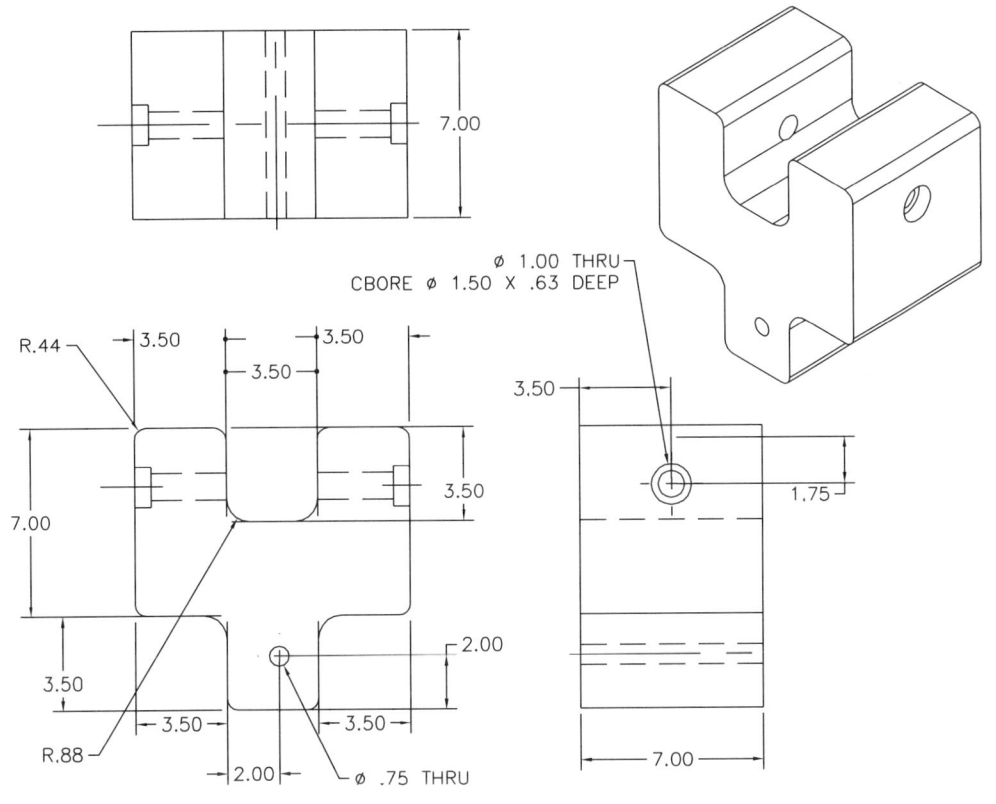

CREATING A NEW PART

As your design progresses you will need to add new parts and components to it. In Mechanical Desktop, these multipart designs are called **assemblies.** You will tell Mechanical Desktop that you want to create a new part, then place it into a catalog when the part is finished.

Use the following steps to create a new part.
Select from the pull-down menu:

Part

➢ Part > New Part

or
Type **AMNEW** ↵
or click the icon.
Type **P** ↵ (*Note:* P is for part.)

Enter the Part Name: <Part 2> (*Type in a part name for this new part.*) ↵
New Part Created

Once parts are created they are placed into an area called the **Assembly Catalog.** Parts can be recalled from the catalog, for use in an assembly (see Chapter 11).

49

If you would like to remove a part from the modeling screen to make room for the new one, use the **ERASE** command, but answer **No** when asked if you want its definition removed (from the catalog). The screen will clear the part; however, it can be redisplayed during assembly creation (see Chapter 11).

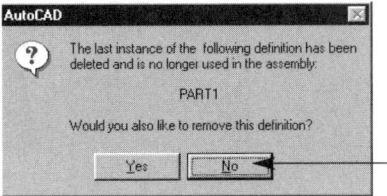

To view the parts you have in the catalog:
Type **AMCATALOG** ↵
or click the Catalog icon.

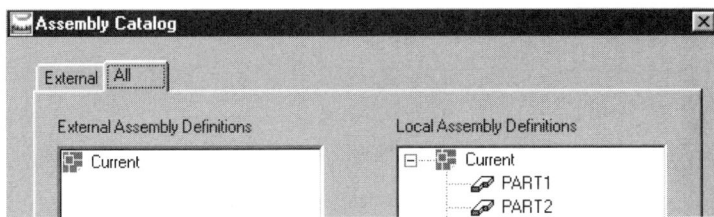

CHANGING WORKING FACES

Once your model has been converted from a 2D profile into 3D feature you will need to work on the other faces to add detail to the model. Once this new face, known as a **sketch plane,** is identified and selected, you can do 2D sketching on the face. Once the sketching is complete, you can apply extruding, revolving, sweeping, or other 3D commands to the model face.

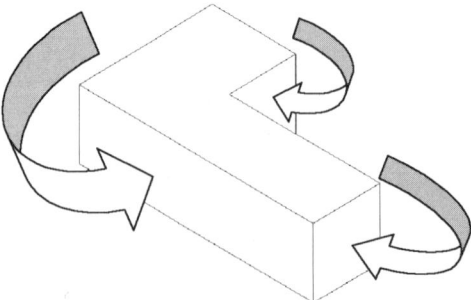

Other model faces to be detailed

CREATING A NEW PART | 51

To change model faces (sketch planes):

Step 1 Type **AMSKPLN** ↵
or click the icon.

Step 2 Click on the face and ↵.

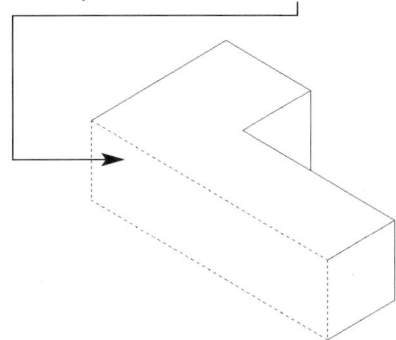

Step 3 Click or toggle the select (left mouse button) to rotate the UCS icon. This icon shows the X, Y, and Z orientation. This step orients the face for sketching.

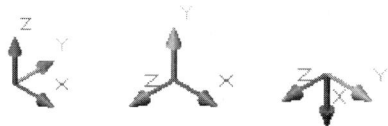

Press ↵ to confirm the orientation desired.
The face (new sketch plane) is ready to be sketched on.

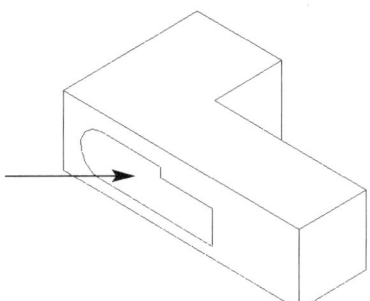

ADDING NEW FEATURES TO A PART

Once you have identified and defined the new face with sketch plane, you can sketch, constrain, and create additional features on the face. You can then appply extruding, revolving, sweeping, lofting, and other 3D features to the constrained sketch on the new part face.

Step 1   Fully constrain the sketch using:

- AMPROFILE

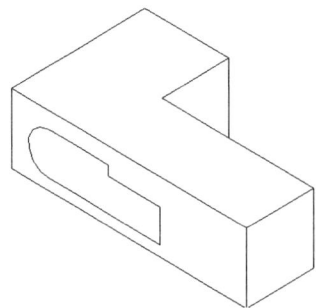

then apply

- AMPARDIM

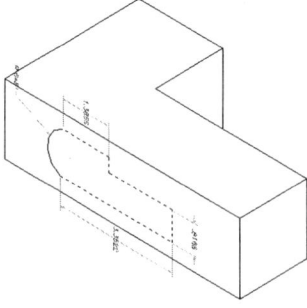

Step 2   For the practice example, let's apply extrude to the new profile.
Type **AMEXTRUDE** ↵
or click the icon.

CREATING A NEW PART | **53**

Select the profile and ↵.

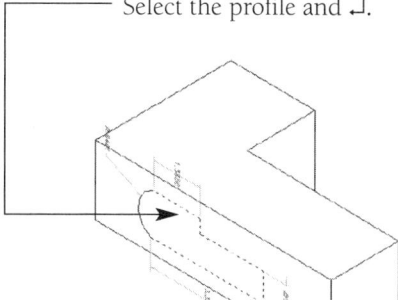

Select the operation and distance, then click **OK**.

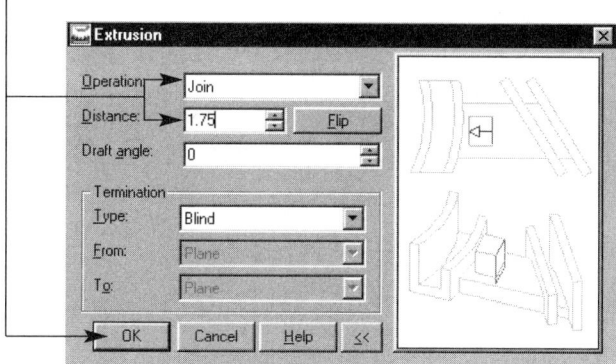

The extrusion is applied to the new face.

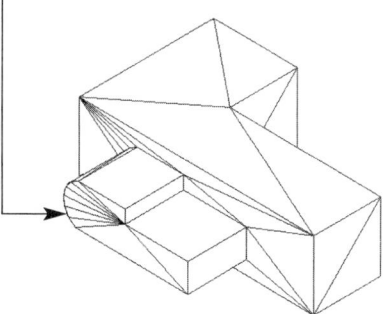

EDITING THE MODEL WITH THE DESKTOP BROWSER

Mechanical Desktop has powerful parametric editing capabilities in a tree type editor known as the **desktop browser.**

Let's look at the desktop browser in the model mode and see what can be edited.

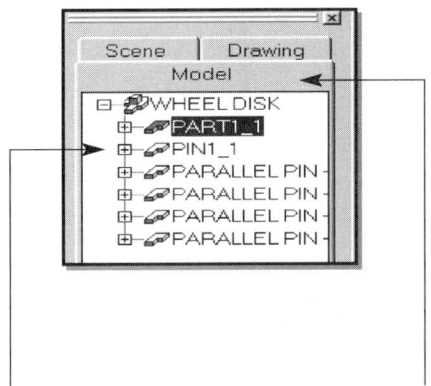

Open any model you have been working on and make sure you are set in the model mode.

Click the **Plus** button to open the details.

Note that when you click on a property in the desktop browser, it is highlighted on the screen.

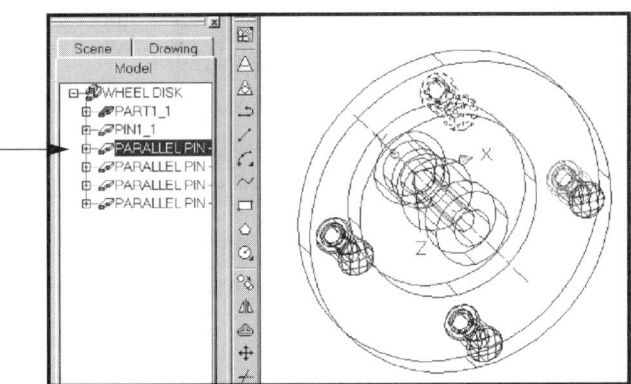

Click on the Plus icon to expand the property options.

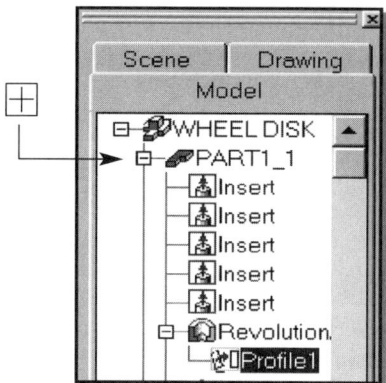

Right-click on a property feature to reveal editable properties. The **Edit** and **Edit Sketch** options allow you to edit down to the sketch level.

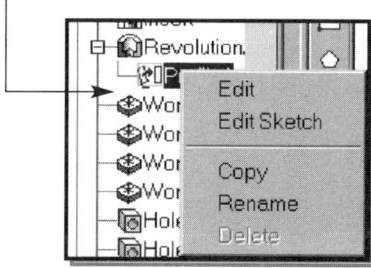

Select one of the 3D-feature part names and right-click on Profile.
Now, try editing with Edit Sketch. Note that the feature reverts to the sketch with constraint dimensions visible.

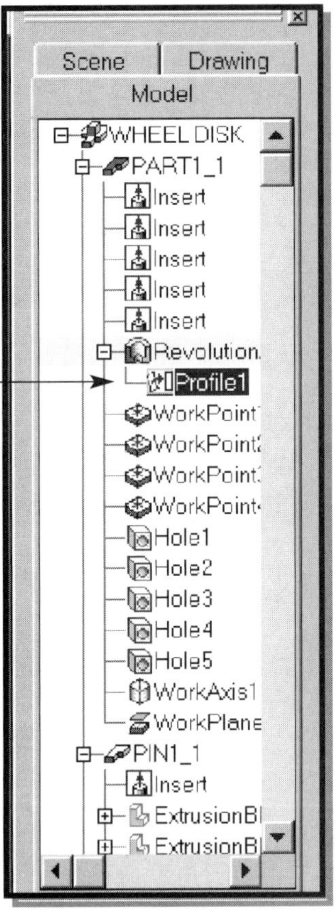

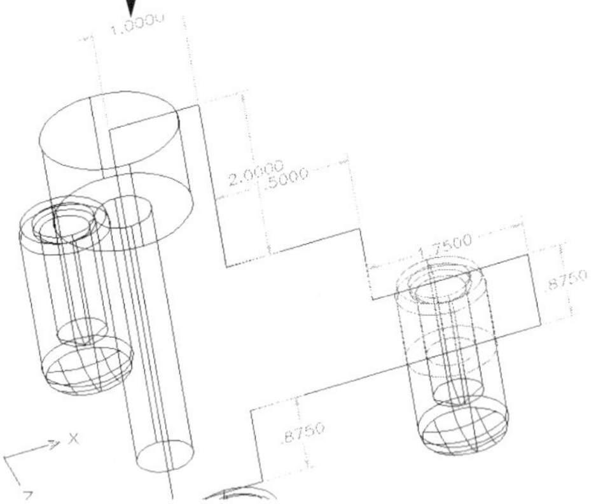

Type **AMMODDIM** or click the icon to modify one of the dimensions on the sketch.
Now, select the dimension you want to modify, key in the new dimension, and press ↵.

Type **AMUPDATE** or click the **Update Part** icon.
The part returns to a 3D feature with the new dimensions.

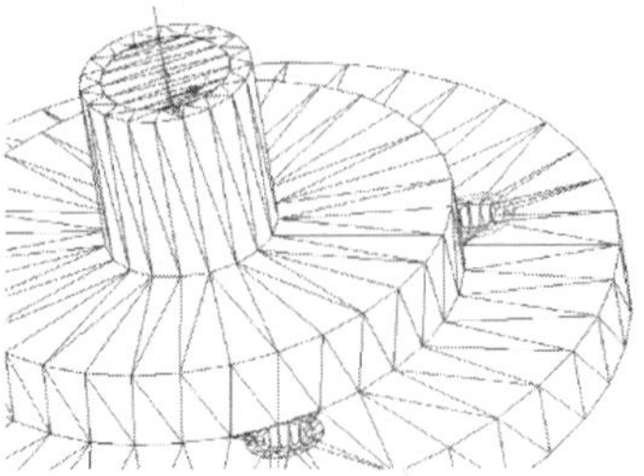

Important: You will note the element highlighted in yellow in the desktop browser if the part has not been updated. This is to notify you to use the update (**AMUPDATE**) to see your changes.

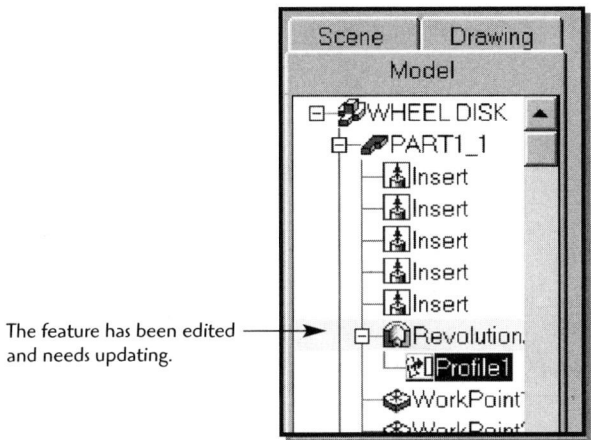

The feature has been edited and needs updating.

THE DIFFERENCE BETWEEN SKETCH PLANES AND WORK PLANES

If you have worked with 3D drawings or solids you are probably familiar with the **UCS** (user coordinate system). The UCS is a way of defining your work surface. So why are both a **sketch plane** and a **work plane** necessary? Mechanical Desktop goes well beyond the capabilities of standard 3D drawing programs and so needs more flexibility for defining 3D space. Sometimes you need a surface plane to reference, but there may be no surface on the current part that can define this plane. In this situation you would define a new work plane. Sketch planes are very similar to the UCS but are easier to define. A sketch plane can be any existing face of a part, or even an existing work plane. When you define a sketch plane, you are saying, "I want to draw or work here." Let's look at both sketch planes and work planes.

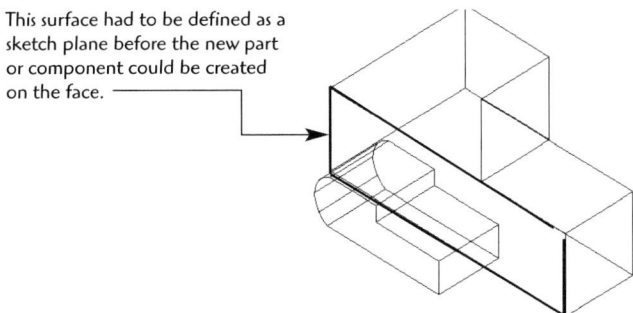

This surface had to be defined as a sketch plane before the new part or component could be created on the face.

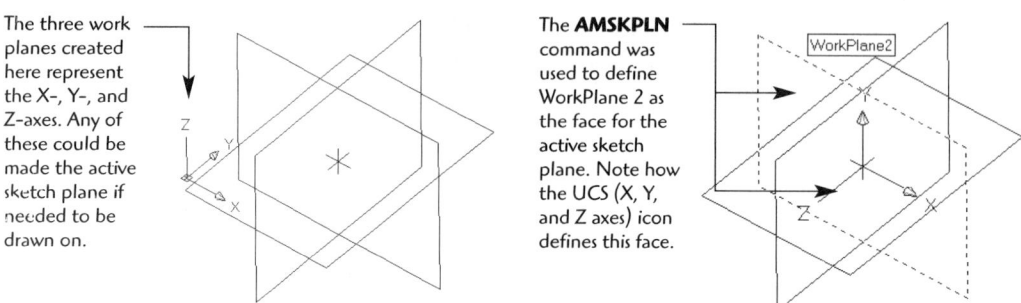

The three work planes created here represent the X-, Y-, and Z-axes. Any of these could be made the active sketch plane if needed to be drawn on.

The **AMSKPLN** command was used to define WorkPlane 2 as the face for the active sketch plane. Note how the UCS (X, Y, and Z axes) icon defines this face.

WORK AXES

Not all reference items are planes. Some are better represented as axes. For example, a round cylinder or a sphere has no flat planes to represent surface areas. These objects are better defined with a **work axis**. Once a work axis is defined, it can be used as a reference point for a work plane or work point, which will be discussed shortly.

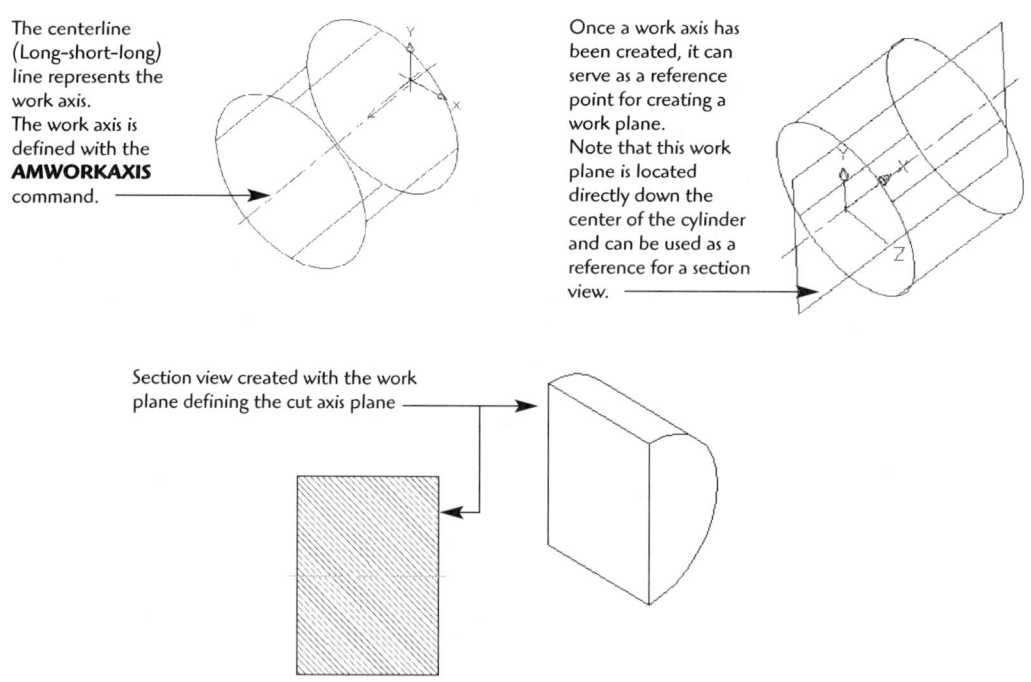

The centerline (Long-short-long) line represents the work axis. The work axis is defined with the **AMWORKAXIS** command.

Once a work axis has been created, it can serve as a reference point for creating a work plane. Note that this work plane is located directly down the center of the cylinder and can be used as a reference for a section view.

Section view created with the work plane defining the cut axis plane

WORK POINTS

Often, you may need to identify points that are not planes or axes. You may use construction lines to get to a point, then mark the point with a **work point**. Work points help identify points such as centers for points, quadrants, or drilled hole locations. Work points can be placed in position with osnaps or just by clicking on a point.

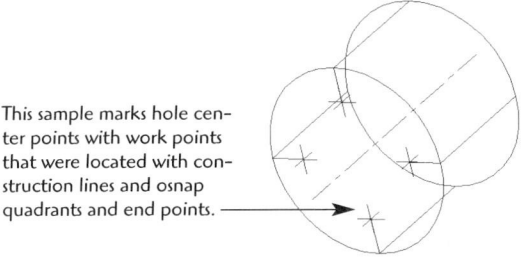

This sample marks hole center points with work points that were located with construction lines and osnap quadrants and end points.

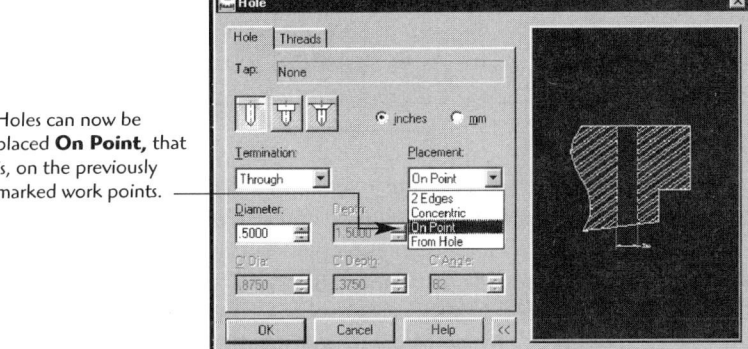

Holes can now be placed **On Point,** that is, on the previously marked work points.

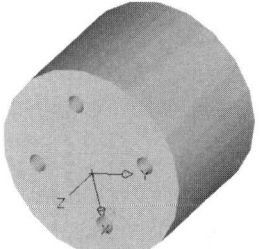

For a Deeper Understanding about
Work Planes

Fully understanding work plane application takes a while, so don't get frustrated with the complexity of this section. You will want to use the section as a reference and return to it in the future. Remember, work planes are created to provide surfaces to reference or to sketch profiles on. Planes can be defined in many different ways.

You learn the different ways of defining work planes as you have the need for them on your model. Let's look at some examples and settings of the Work Plane dialog box.

CHAPTER 4

Two modifiers are set to define a work plane.

The UCS or World coordinates can also be used to define a work plane.

The **Create Sketch Plane** setting is usually checked. This automatically sets the work plane as the active sketch plane.

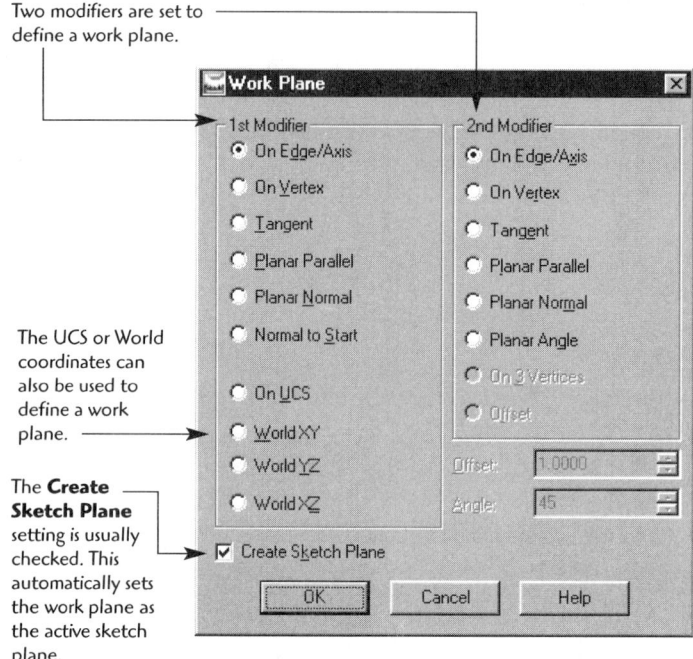

Type AMWORKPLN ↵. (The Work Plane dialog box appears.)

When you define a new work plane you need to define at least two modifier elements such as edges, vertex, tangent points, or existing planes. Even if there are no elements on the modeling page yet, an X, Y coordinate can be assigned as a starting point. If you are familiar with creating planes with the UCS system, you can select **On UCS** to create the new work plane. The world coordinate is the default plane at the startup modeling screen. If you needed a work plane 90° and down from the world coordinate, you would select **World XZ**. If you needed a work plane 90° and to the side, you would select **World YZ**. For a work plane "straight on," you would select **World XY.**

The **AMBASICPLANES** command will also create the three basic X, Y, and Z work planes all in one step.

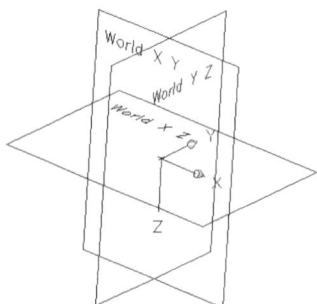

Let's look at some useful applications for work planes. What if you need to place a hole in an L-plate at an angle perpendicular to the two top edges? Here is one approach using a work plane/sketch plane.

- On Edge/Axis
- On Edge/Axis

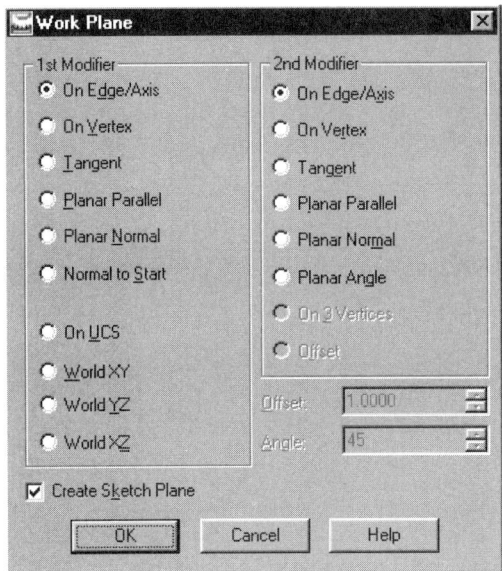

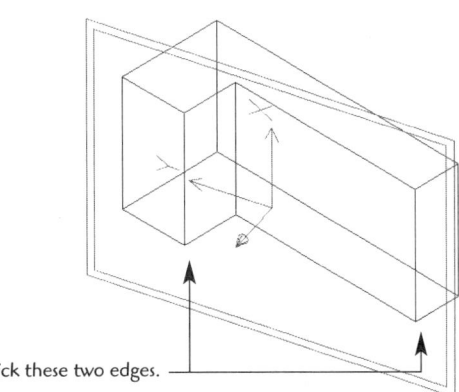

Pick these two edges.

A construction line was placed on the new work plane, and a work point was placed at the endpoint of the line. The drilled hole placement was set to On Point.

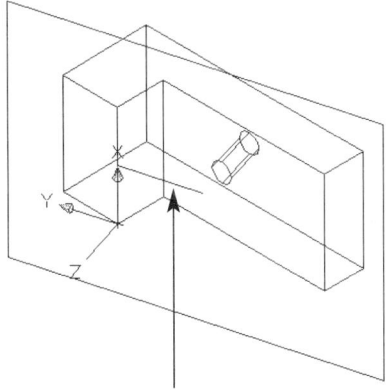

The construction line is drawn on the new work plane. A work point is placed on the endpoint of the construction line.

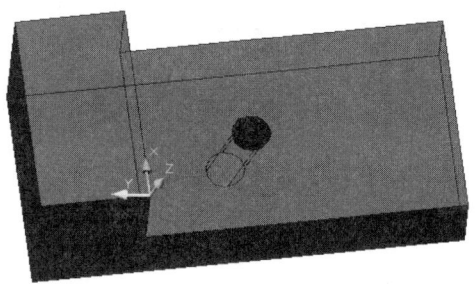

The angled hole placement applied

What if you need to drill and tap a hole on the tangent surface of a cylinder? A work plane would be helpful in this case. We will proceed by first creating a work plane through the center of the cylinder, then offsetting a work plane to the tangent side.

First, create a work axis down the center using **AMWORKAXIS**.

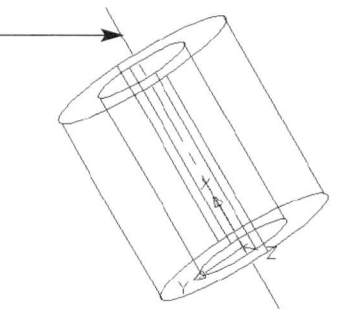

Then, create a work plane in the middle of the cylinder. Type AMWORKPLN to display the work plane dialog box.

- On Edge/Axis
- Planar Angle

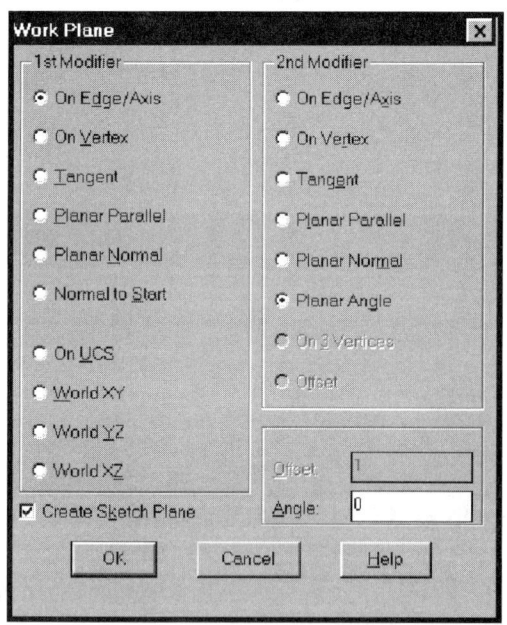

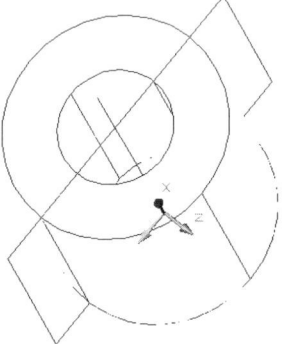

After clicking **OK**, pick the work axis, then type **Y** ↵↵↵.

(*Note:* Typing the Y simply gives Mechanical Desktop an axis start point. The three "Enters" accept any defaults and will place the new work plane in the center of the cylinder.)

CREATING A NEW PART | 63

Now we will use the Planar Parallel, and Tangent to create the tangent work plane.

- Planar Parallel
- Tangent

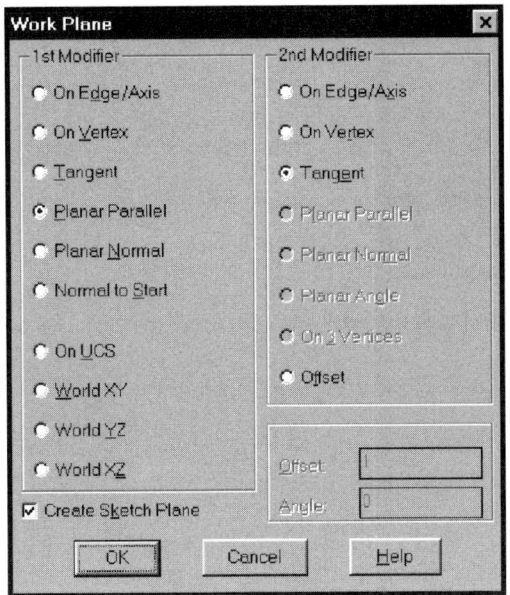

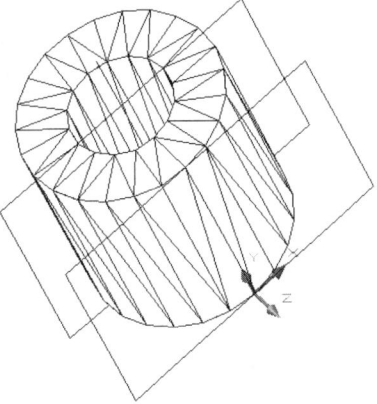

A construction line is added from a quadrant point, and a work point is added to its endpoint.

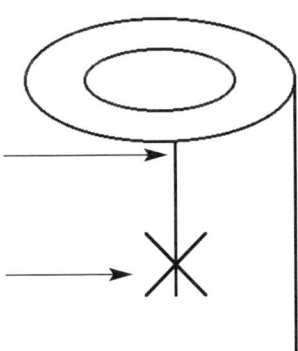

Using the **AMHOLE** command, set the placement to On Point.

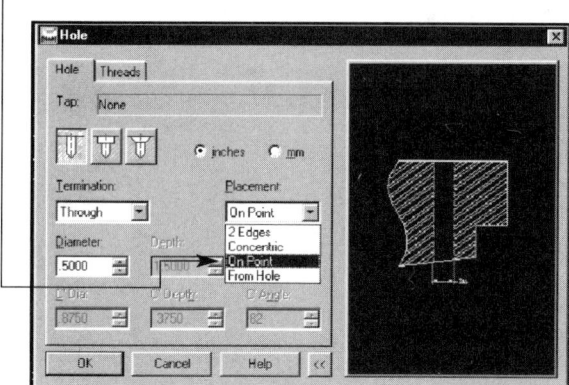

The cylinder can now be drilled from the tangent side.

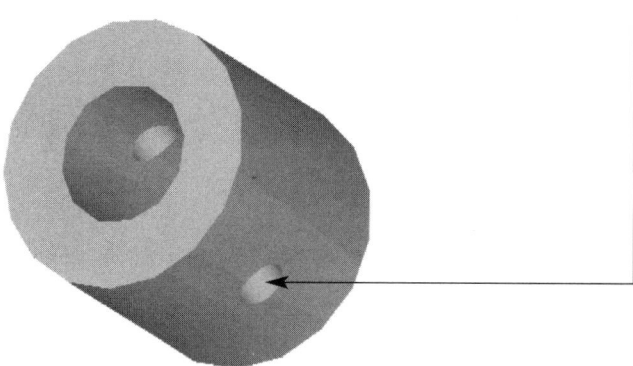

Now that a few work planes have been created on the cylinder, let's look at some additional work planes that can be added from this configuration.

The two modifiers are represented by the bullets:

- On Edge/Axis
- Planar **Normal**

The new work plane is "normal to" or at a right angle to the original work plane.

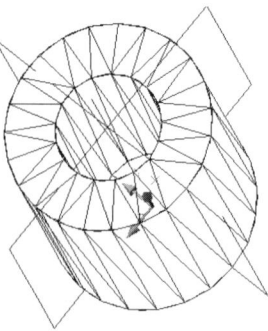

- On Edge/Axis
- Planar **Angle**

The angle is set to 45° here, but could be set to any angle.

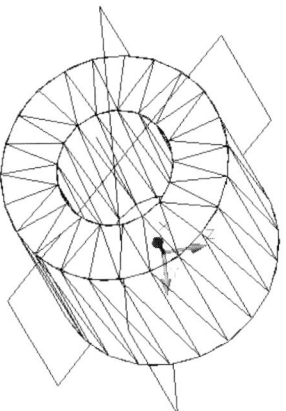

- On Edge/Axis
- Planar **Parallel**

The work plane is parallel to the edge of the part.

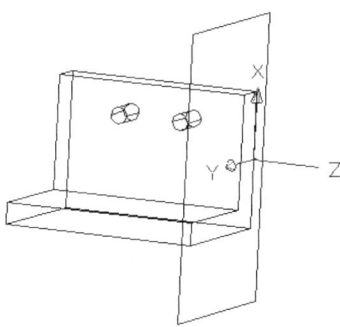

- Planar Parallel
- **Offset**

The new work plane is offset from the parallel plane.

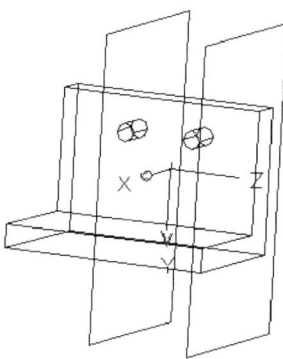

- On Edge/Axis
- On Edge/Axis

Two work axes are placed through the center of the holes, providing two axis modifiers for the work plane.

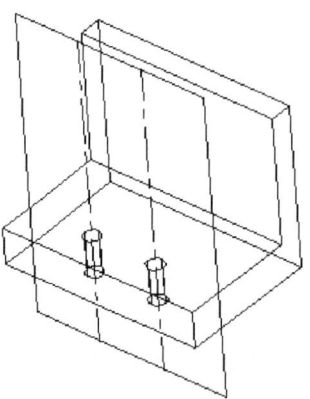

- On Vertex
- On 3 Vertices

The 3 Vertices modifier allows you to select corners to produce an angular work plane and the final shape cut block.

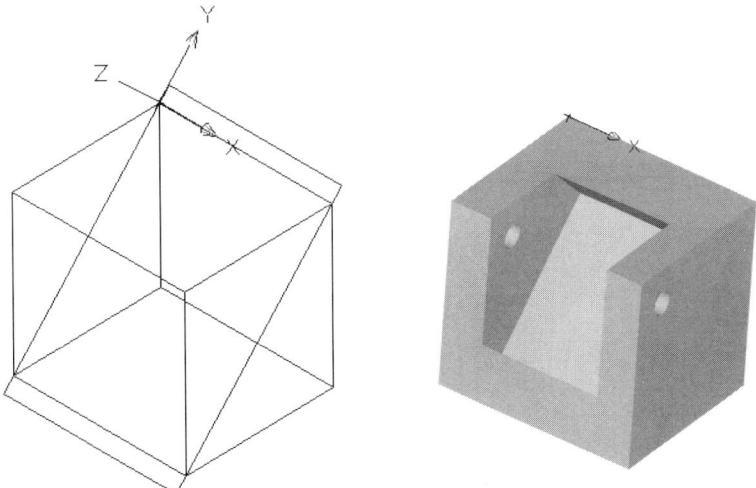

In addition to typing in the commands, you can use one of the icon buttons. Here are some of the icons for work planes, work points, and work axis.

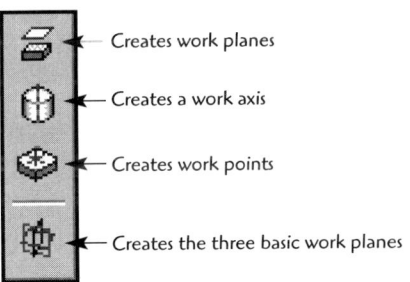

Creates work planes

Creates a work axis

Creates work points

Creates the three basic work planes

Drawing Project 4–1

Cleat Post

Model the Cleat to the exact dimensions indicated. Add the fillets and holes, and change sketch planes as needed to complete the model. When you have finished, save the drawing and click **8** for an isometric view. Print out a copy for your instructor.

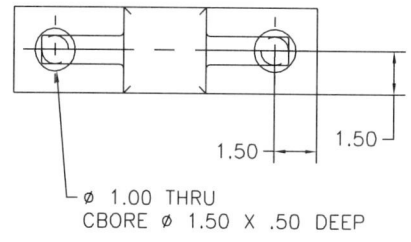

Ø 1.00 THRU
CBORE Ø 1.50 X .50 DEEP

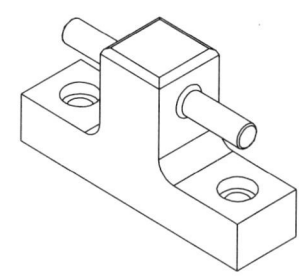

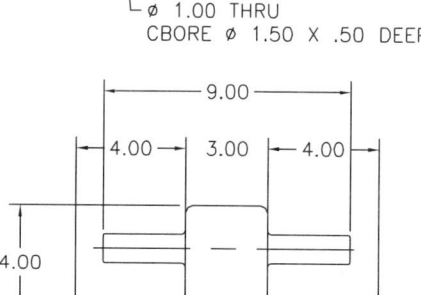

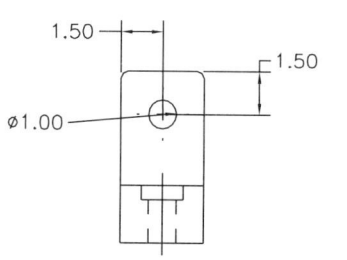

Drawing Project 4–2

Slot Bracket

Model the slot bracket to the exact dimensions indicated. Add the fillets and holes, and change sketch planes as needed to complete the model. When you have finished, save the drawing and click **8** for an isometric view. Print out a copy for your instructor.

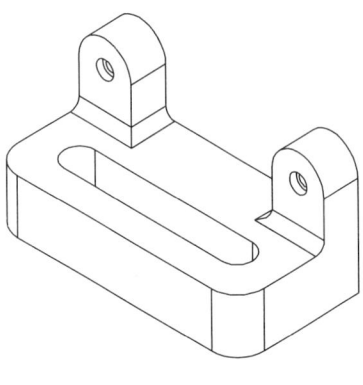

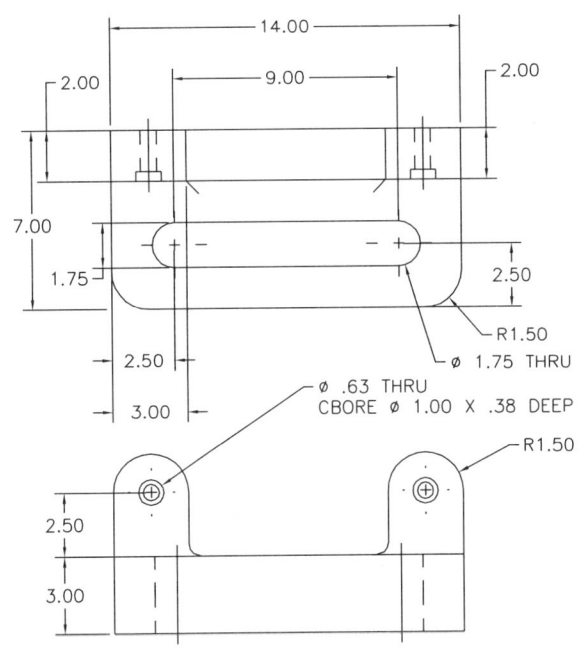

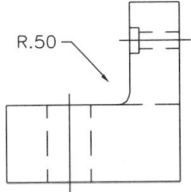

Drawing Project 4–3

Adjustment Plate

Model the adjustment plate to the exact dimensions indicated. Add the fillets and holes, and change sketch planes as needed to complete the model. When you have finished, save the drawing and click **8** for an isometric view. Print out a copy for your instructor.

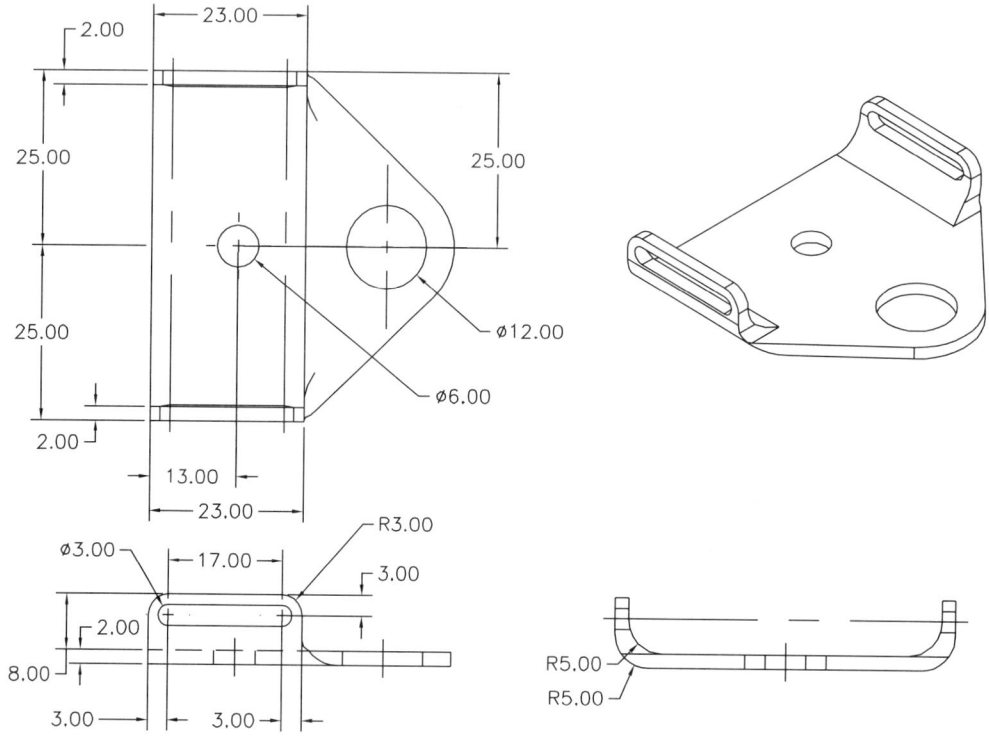

5

REVOLVING

Once a sketch has been fully constrained, a variety of 3D features can be applied. We have already seen the application of the Extrude command, now let's see how other 3D feature–creation commands turn your sketch into features. The **Revolve** command is a powerful feature-creation command. Let's see how it is applied.

Step 1 Start by Creating the sketch. (*Note:* You should plan your sketch as a half-section of the final shape.)

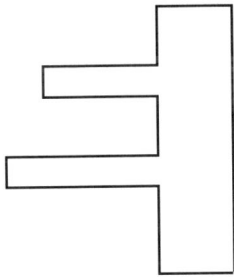

Step 2 Profile your sketch.
Type **AMPROFILE** ↵
or click the icon.
Select the profile and ↵.

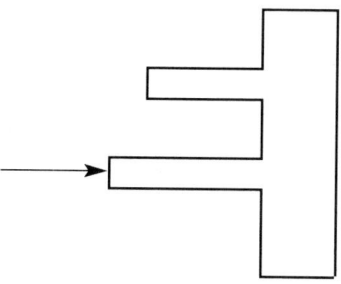

Step 3 Type **AMPARDIM** ↵
or click the icon.
Click on a side, then away from the side.
The constraining dimension is added.

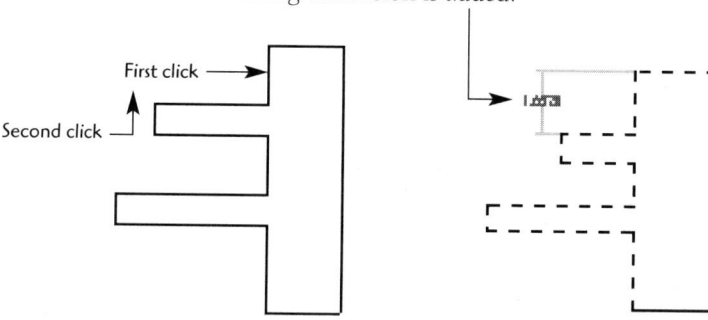

Continue until the sketched profile is fully constrained.

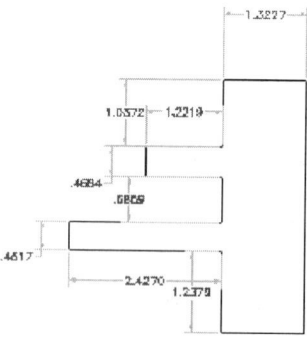

Step 4 The Revolving of the fully constrained sketch...
Type **AMREVOLVE** ↵
or click the icon.

Select the **revolution axis**. (*Hint:* Think of the revolution axis as the point to spin around.)

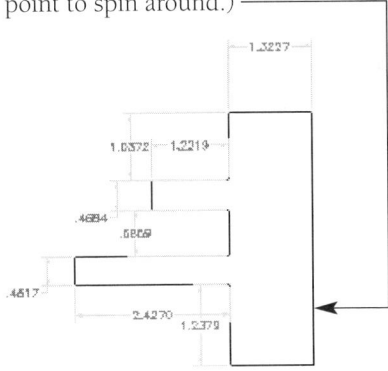

In this case leave the default settings, and revolve a full 360°. Click **OK**.

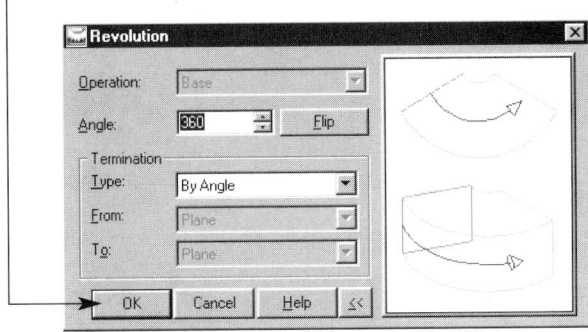

The Revolved profile is now a 3D feature.

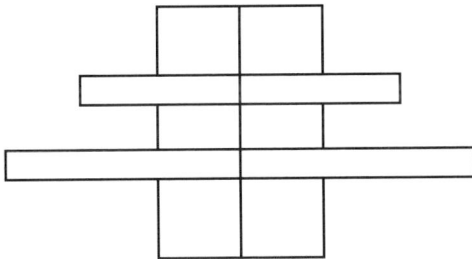

Use the 3D Orbit tool to rotate the part into a better view.

For a Deeper Understanding about
Revolving

Once the 2D sketch is completed, switching to an isometric, or iso, view helps you see the revolved feature. To quickly display sketch or feature in an iso view, simply type the **8** and ↵.

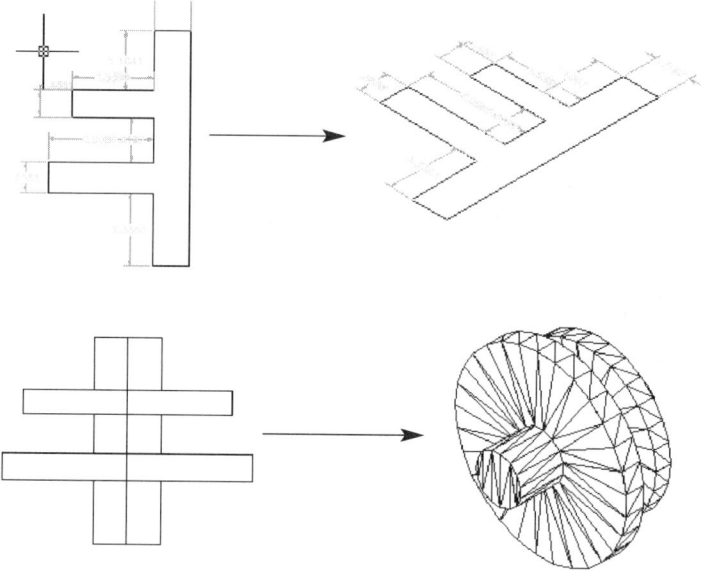

Now let's look at the options for revolving profiles.

The angle, or the amount of revolution, can be adjusted.

The termination of the revolution can be from the **midplane**, from a **defined plane**, or **from-to** a point or surface.

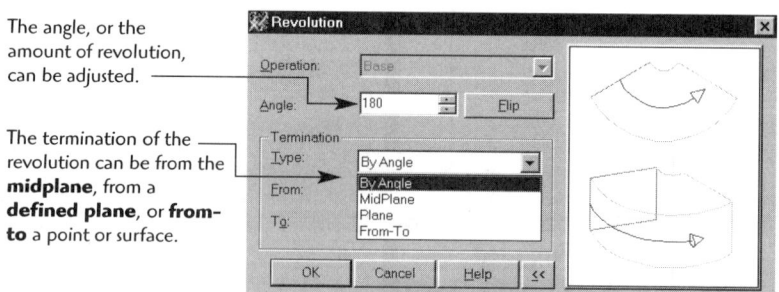

REVOLVING | 75

Examples:

By Angle (in this case 180°)

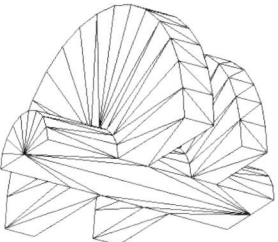

From the MidPlane (Note that the profile revolves in both directions from the profile.)

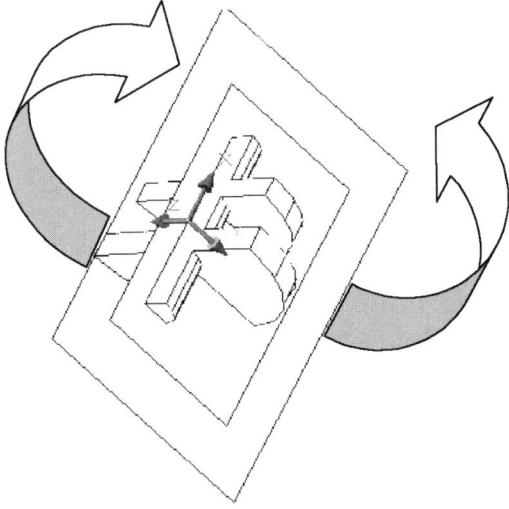

From-To (from a start plane to an end plane)

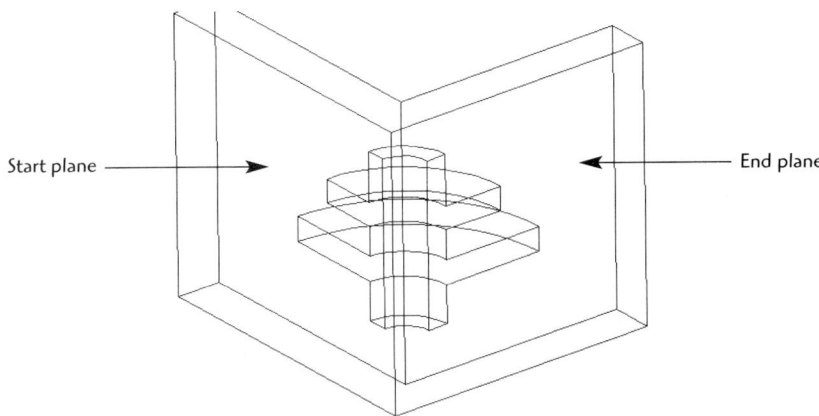

Start plane — End plane

When you use the Operation: **Join** option, you will note an extended series of revolution options. Planes and faces can be used as points to accurately terminate a revolution.

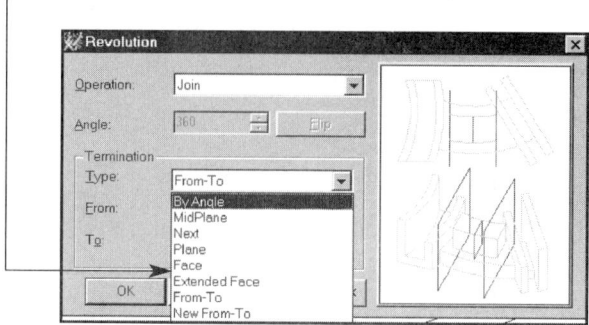

When you use the Join option both the original base shape and the revolved profile shape become one.

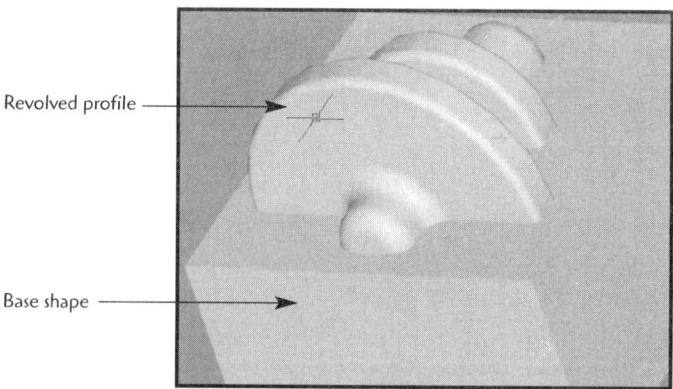

Revolved profile

Base shape

Unique shaped cavities can be cut into a base part with the Operation: Cut option.
Once a base part is created, a new sketch plane is created on a face or surface of the profile. The sketch is fully constrained and Revolution is selected. In the Operation: option, Cut is selected.

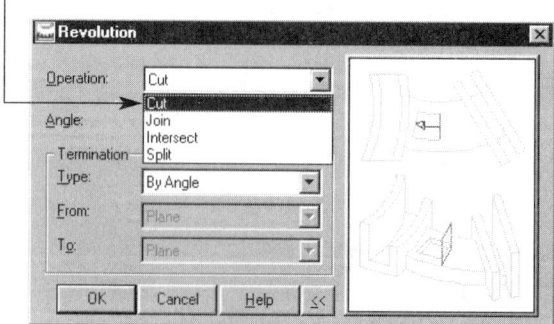

You can create pockets, cavities, mold, or die shapes using the **Cut** option of revolution.

Rendering of a revolved cut profile

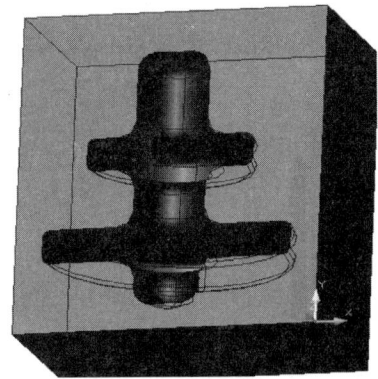
This view is set with **Shaded View** and **Edges on.**

The Operation: **Intersect** option will produce a unique solid out of the area shared by the original feature and the revolved profile.

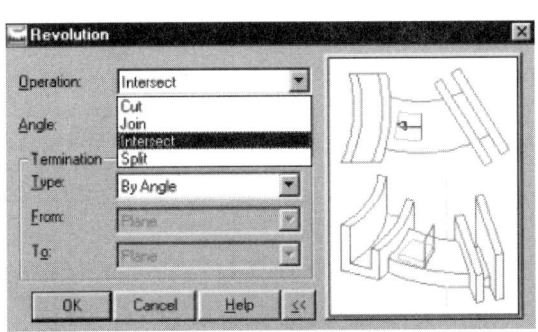

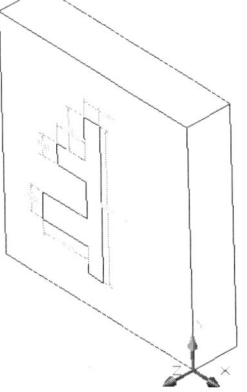

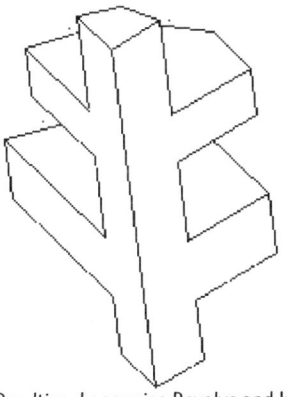
Resulting shape using Revolve and Intersect

Using the **Split** option of revolution creates both the cavity and a solid of the revolved shape. This option is similar to Intersect, but both parts remain.

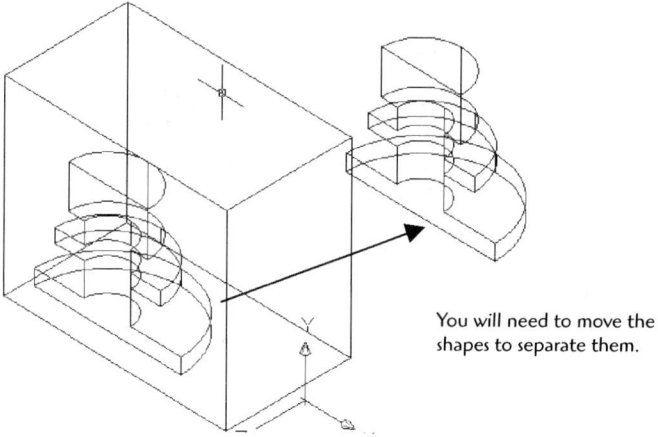

You will need to move the shapes to separate them.

Drawing Project 5–1

Revolve Pulley

Model the revolve pulley to the exact dimensions indicated. Use Revolve and add the fillets and holes as needed to complete the model. When you have finished, save the drawing and click **8** for an isometric view. Print out a copy for your instructor.

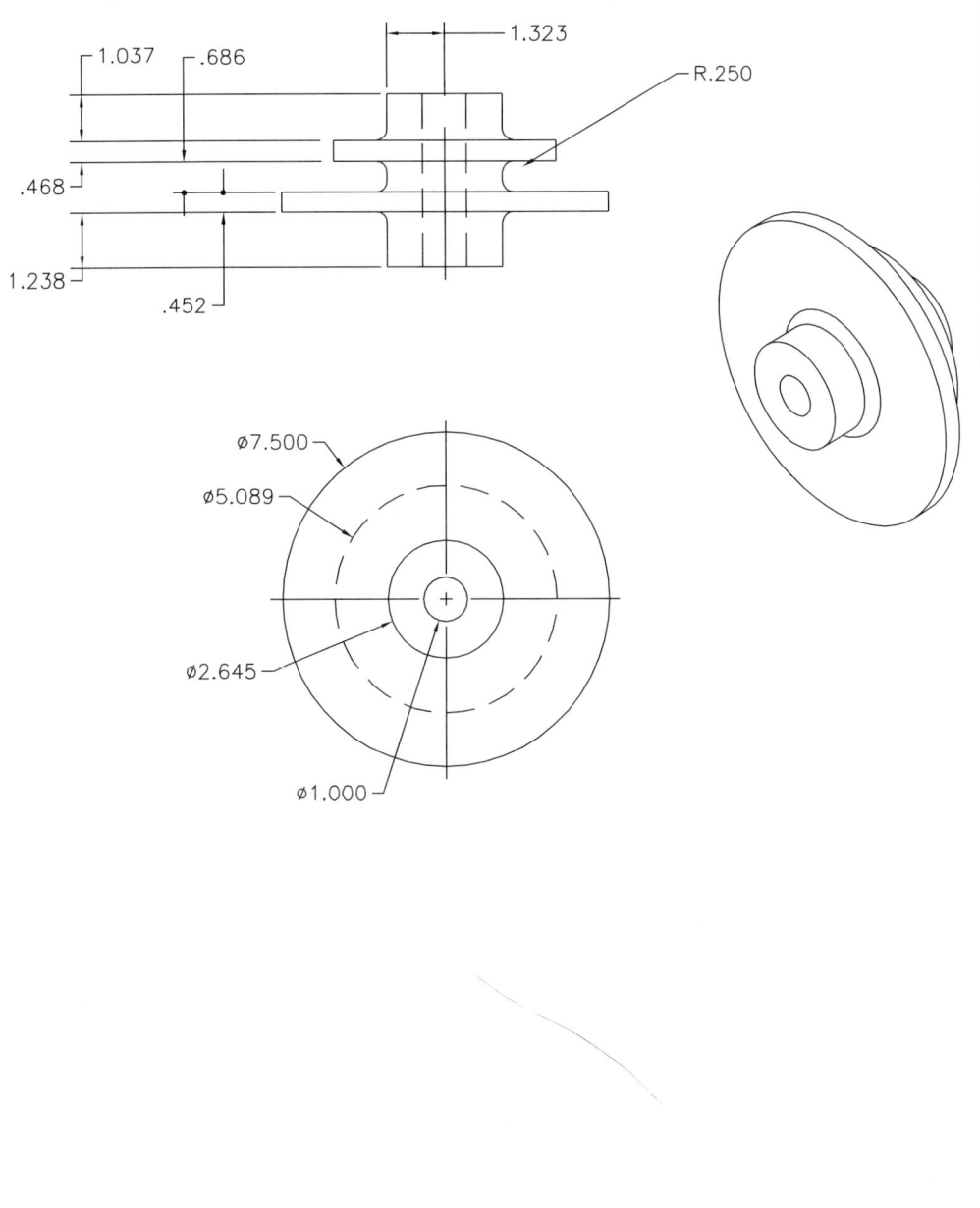

Drawing Project 5-2

Shift Spool Valve

Model the shift spool valve to the exact dimensions indicated. Use Revolve and add the fillets and holes as needed to complete the model. When you have finished, save the drawing and click **8** for an isometric view. Print out a copy for your instructor.

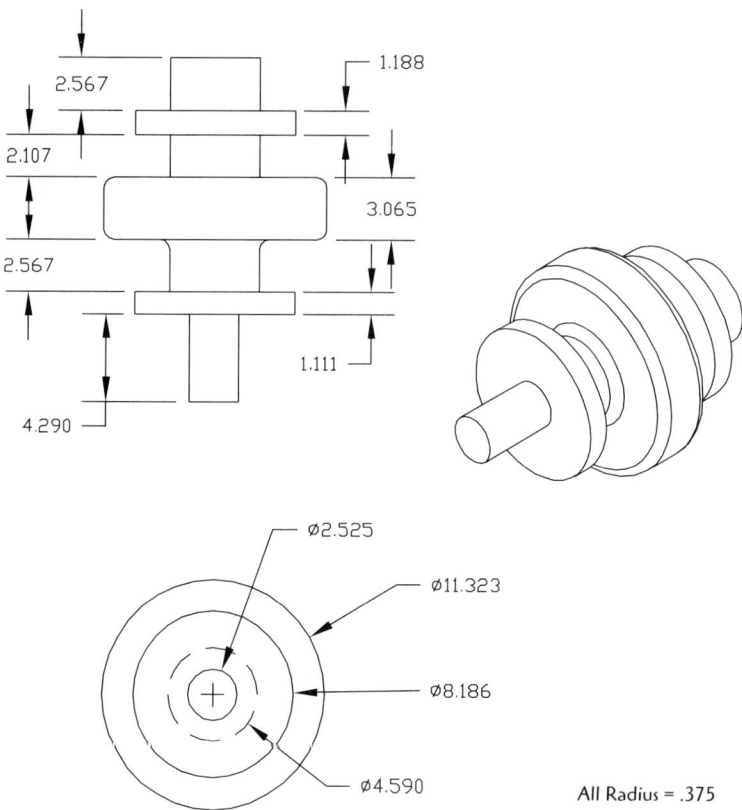

All Radius = .375

Drawing Project 5–3

Flow Throttle Spool

Model the flow throttle spool to the exact dimensions indicated. Use Revolve and add the fillets and holes as needed to complete the model. When you have finished, save the drawing and click **8** for an isometric view. Print out a copy for your instructor.

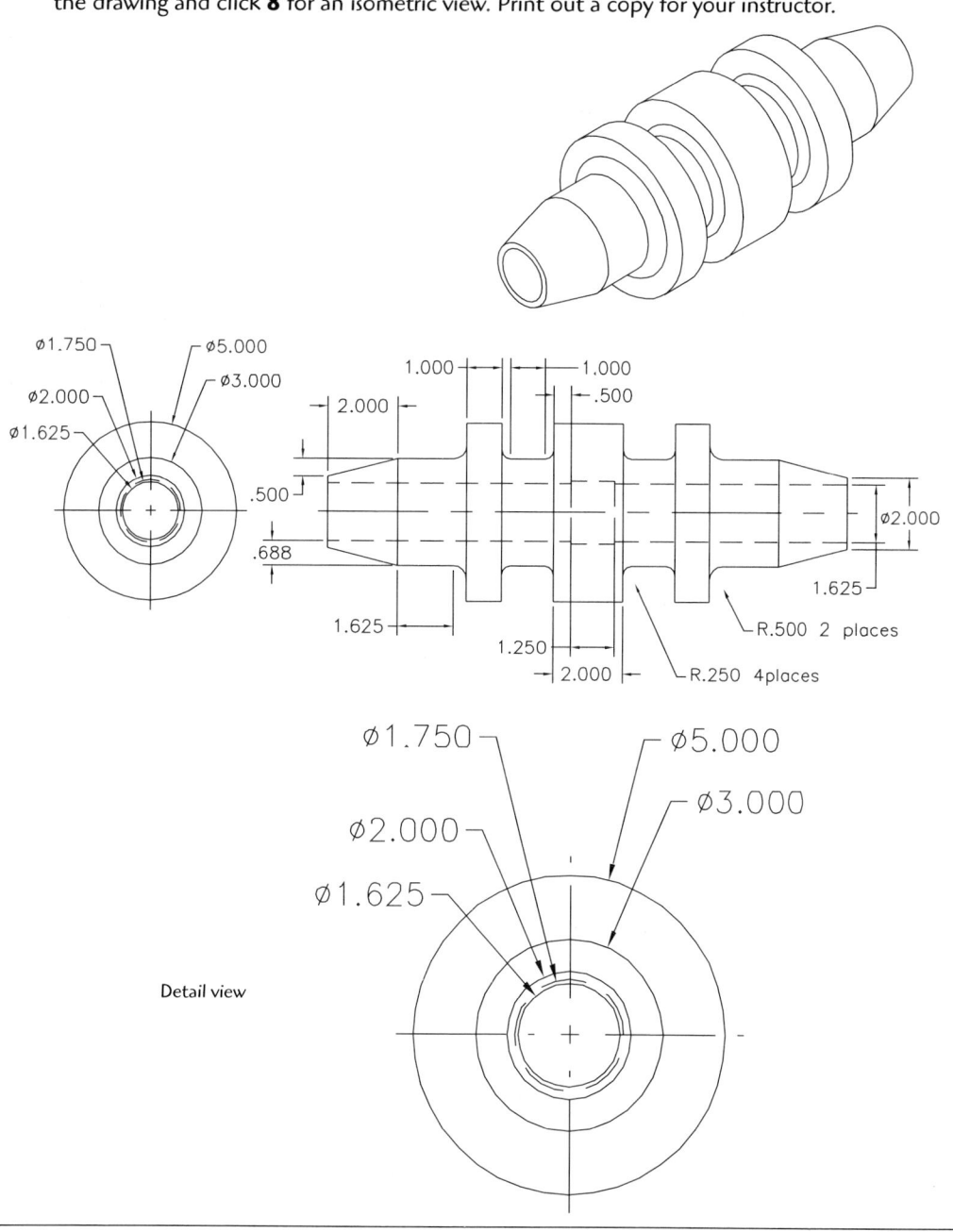

Detail view

Drawing Project 5-4

Micrometer

The specifications shown for the micrometer are incomplete. You must complete the model with your own design to complete the project. You may extract dimensions from a 1" micrometer to help you complete your design. Use revolve, extrude, and any other commands you have learned to complete the model. Take notes and record dimensions on the following sketch. When you have finished, save the drawing and click 8 for an isometric view. Print out a copy for your instructor.

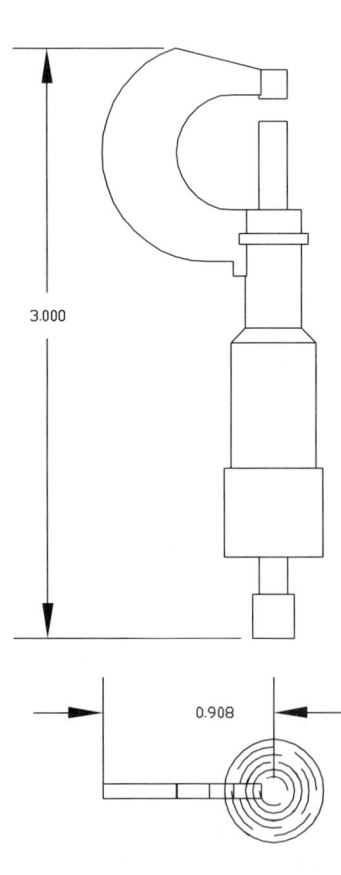
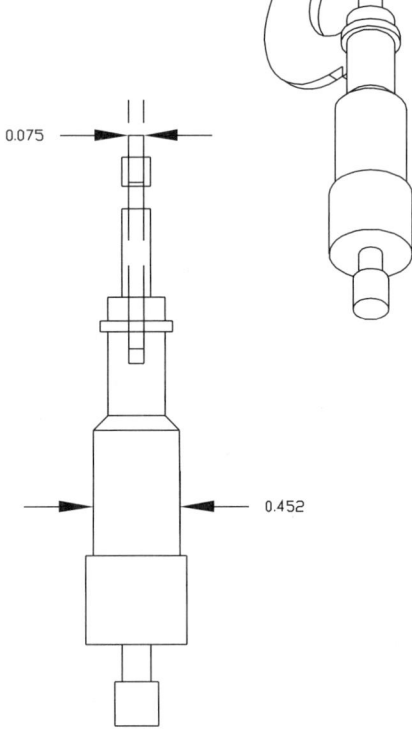

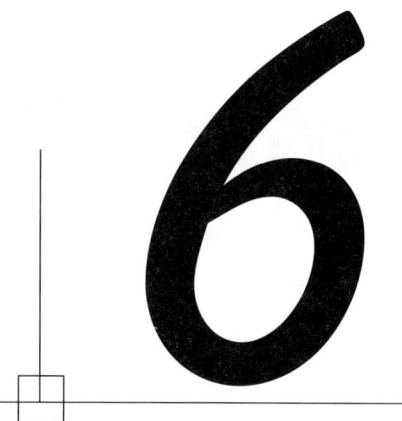

SWEEPING

The process of sweeping is useful for creating parts and components that maintain a constant cross section through a curved path. The process requires you to create a **sweep path** and a **cross section**. Both are fully constrained—the cross section as a profile, and the sweep path as a path. Once fully constrained with dimensions, the cross section is **swept** along the path.

Steps for Sweeping

Step 1 Create a sweep path using a **pline**. (*Note:* Unlike a profile, a path can be open.)
Add standard fillets as necessary to refine the path.
Click **8** to display an Iso view.

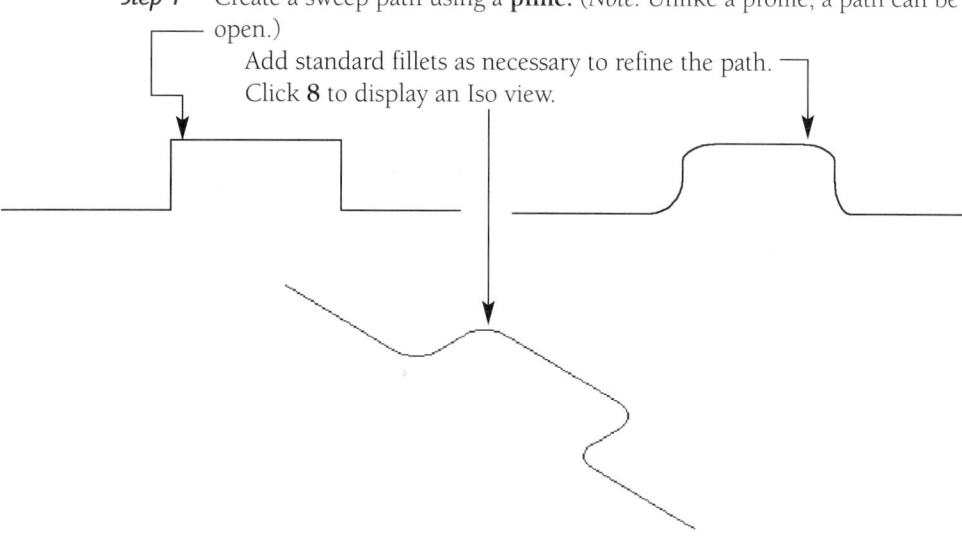

83

CHAPTER 6

Step 2 Constrain the sweep path:
Type **AM2DPATH** ↵
or click the icon.
You will be prompted to

`Pick the Sweep Path` *(Select anywhere on the path.)*

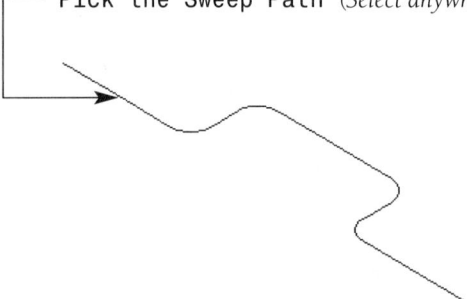

`Select start point of path:` *(Select one of the ends where the profile will be placed.)*

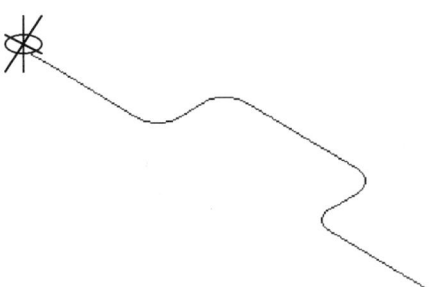

Step 3 Constrain the path with dimensions:
Type **AMPARDIM** ↵
or click the icon.
(Click in the constraining dimensions.)

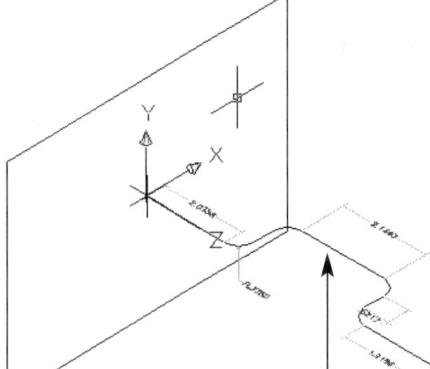

Create a profile plane perpendicular to the path? *(Press ⏎ to accept Yes.)*

Select edge to align X axis: *(Press ⏎ to accept the default alignment.)*

Note the sketch plane created at the end of the path so you can create the profile.

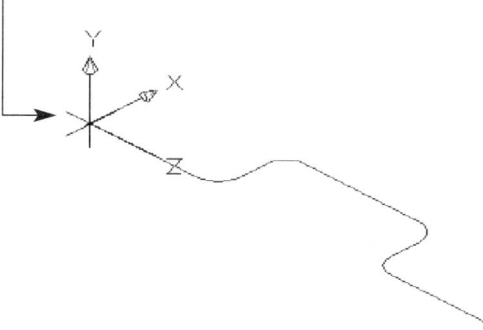

Step 4 Sketch the profile on the end of the sweep path.
(*Hint:* Use the Osnaps to ensure that you locate the profile correctly on the path.)

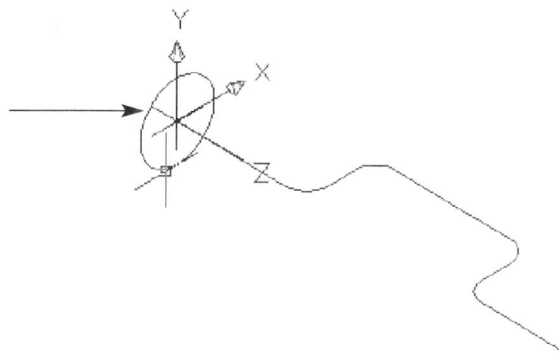

Step 5 Constrain the profile:
Type **AMPROFILE** ⏎
or click the icon.

Select the profile and ↵.

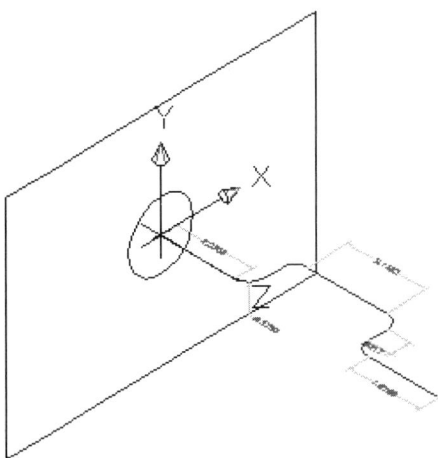

Step 6 Constrain the profile with constraining dimensions:
Type **AMPARDIM** ↵
or click the icon.

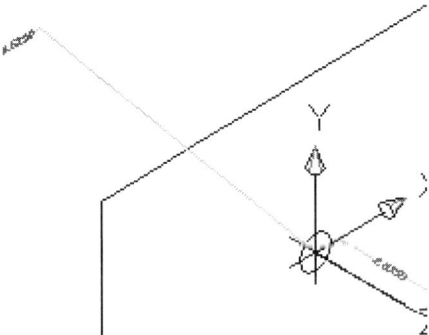

Step 7 Now you are ready for the sweeping!
Type **AMSWEEP** ↵
or click the icon.

Accept the defaults and click **OK**.
The sweep will take a few seconds to generate.

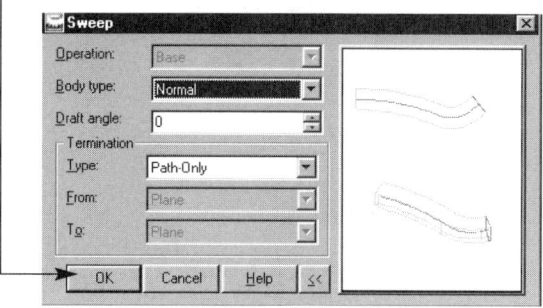

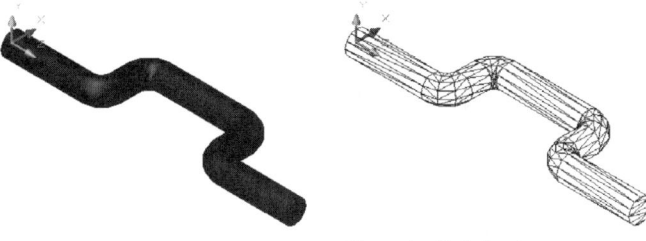

View using Hide Display   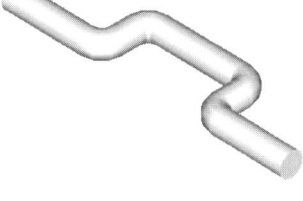

Fully rendered view

For a Deeper Understanding about
Sweeping

Some of the sweep options give you the ability to create unique shapes. Try setting the draft angle to other than 0 and note the interesting shapes.

The draft angle is set to -4 to create a tapered shape.

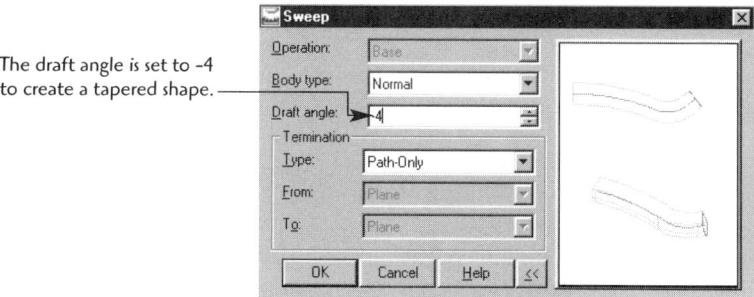

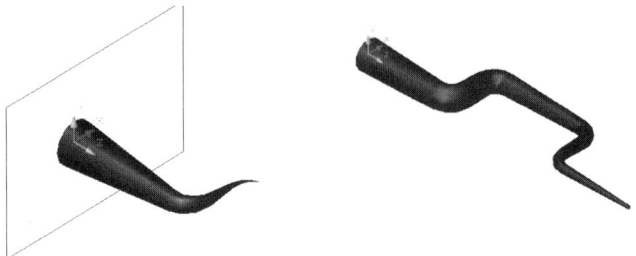

The draft angle is set to -4 (left) and to -2 (right) for unique shape creating

Other options include limiting the extent of the sweep.

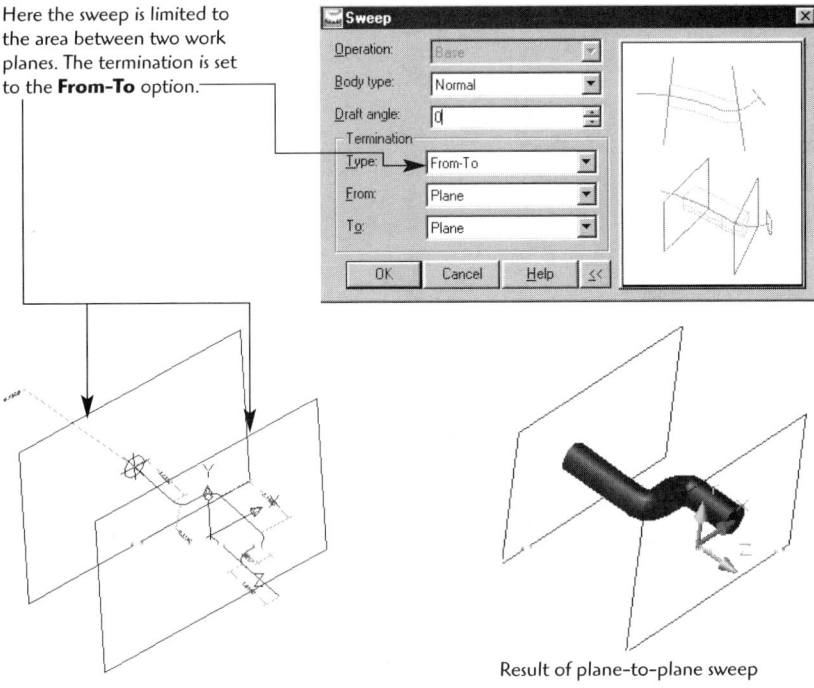

Here the sweep is limited to the area between two work planes. The termination is set to the **From-To** option.

Result of plane-to-plane sweep

LOFTING

The term *lofting* comes from the old ship building trade practice of "lifting off" or lofting rib cross sections from the previous cross section to create the hull profile. The process of lofting is useful for creating parts and components that have a varying cross section through a distance. The process requires you to create at least two or more cross sections. All cross sections are fully constrained. Once fully constrained with dimensions, the cross sections are lofted to create the solid shape.

Steps for Lofting:

Step 1 Create a series of work planes on which to sketch each cross section.
(Click **8** to obtain an Iso view.)

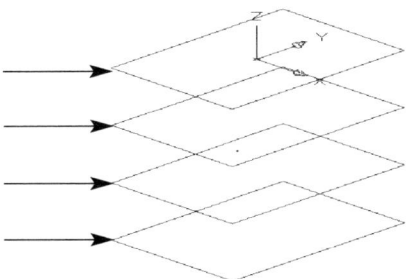

Step 2 Set the sketch plane onto the first work plane.
Type **AMSKPLN** ↵
or click the icon.
Click on the first work plane to set it as the active sketch plane.

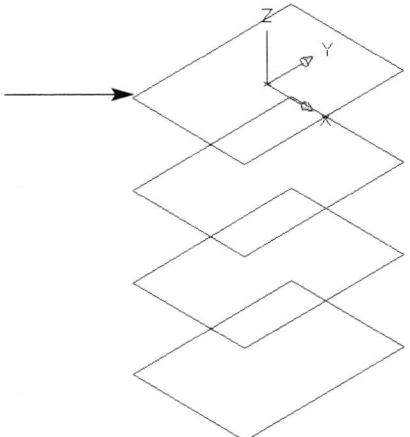

Step 3 Sketch the cross section (profile) on the work plane/sketch plane.

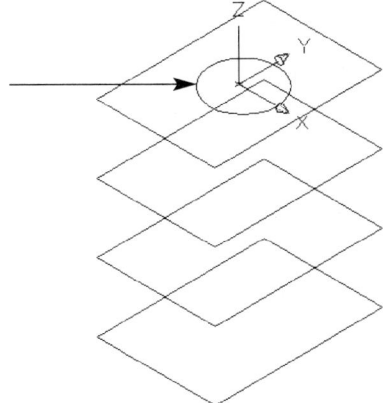

Step 4 Fully constrain the cross section (profile).
Type **AMPROFILE** ↵
or click the icon.
Select the profile and ↵.
Constrain the profile with constraining dimensions.
Type **AMPARDIM** ↵
or click the icon.
Click on the profile and click away to place the constraining dimension.

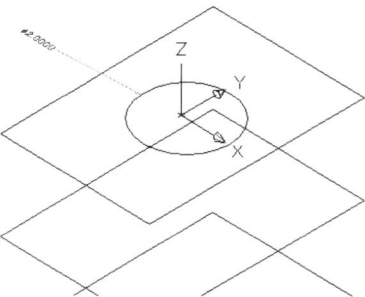

Step 5 Continue steps 2 through 4 for each work plane:
- Activate the work plane to the active sketch plane.
- Sketch the cross section profile.
- Fully constrain.

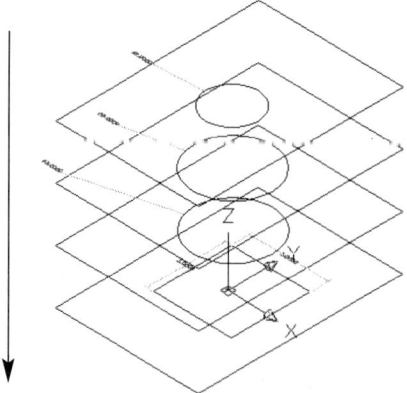

Step 6 You are now ready for lofting.
Type **AMLOFT** ↵
or click the icon.

SWEEPING | 91

`Select profiles or planar faces to loft:` *(This is asking you to select the cross sections.) Starting with the first cross section, select **all** the sections.*

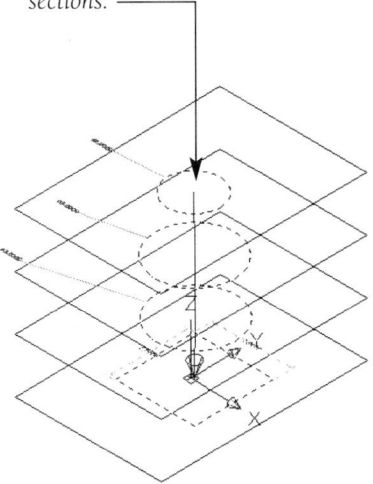

The Loft dialog box appears. Accept the default settings with Type set to **Cubic**, and Termination set to **Sections**. Click **OK**.

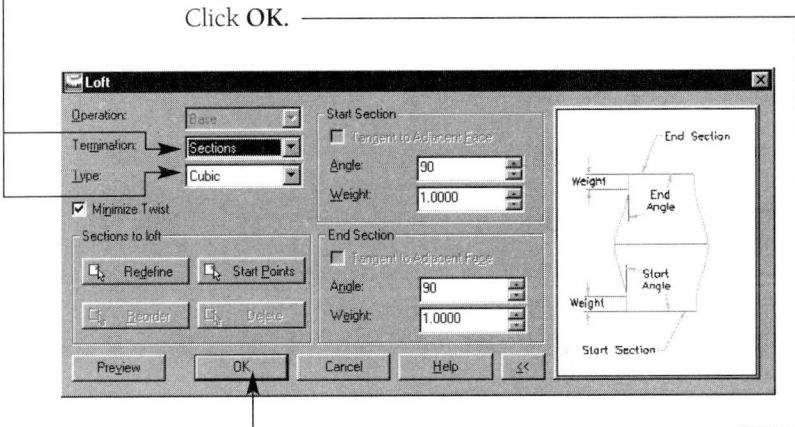

Hint: If the loft fails, try again but click the **Start Points** option and select the first cross section, then OK.

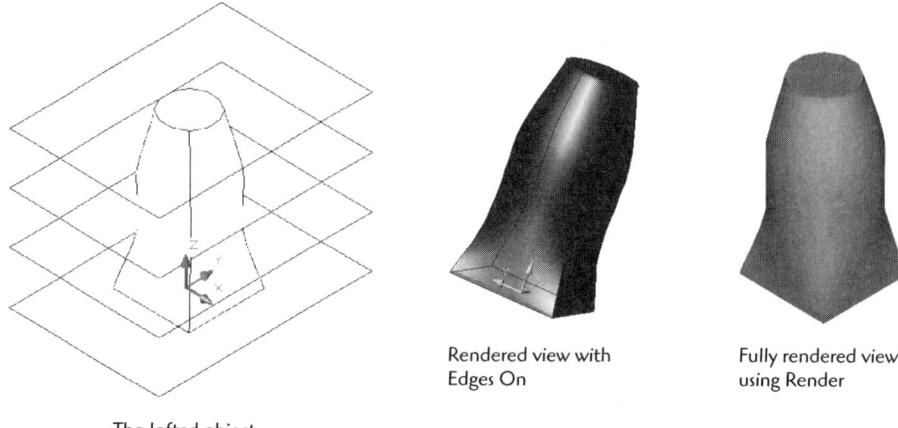

Rendered view with Edges On

Fully rendered view using Render

The lofted object

FIXING LOFTING PROBLEMS

If your lofted object does not loft exactly as desired, there are some adjustments you can make. Occasionally your loft may look twisted. This may mean that your start points are not aligned correctly. To fix this problem, follow this procedure.

Right-click on the **Loft** branch in the desktop browser, then click **Edit**.

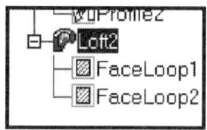

The Loft dialog box appears.
Click on **Start Points**.

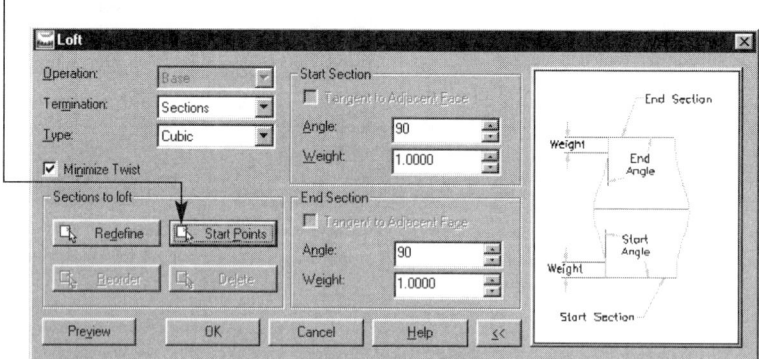

Click a new start point on the cross section profile so the start point icons align.

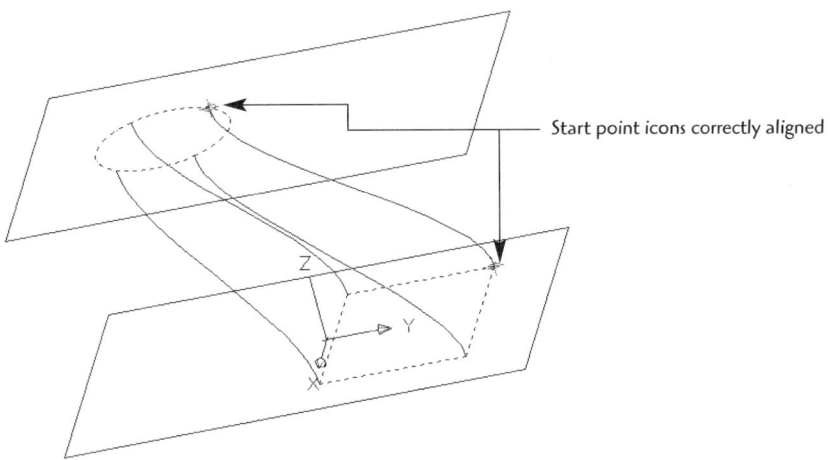

Start point icons correctly aligned

For a Deeper Understanding about
Lofting

Let's look at the types of lofting options:

- Linear
- Cubic
- Closed Cubic

If you are using only two cross section profiles, you can use the **Linear** option. The linear loft will produce a straight, or linear, path between cross section profiles. You will note that the edges are abrupt and do not curve the transition between sections.

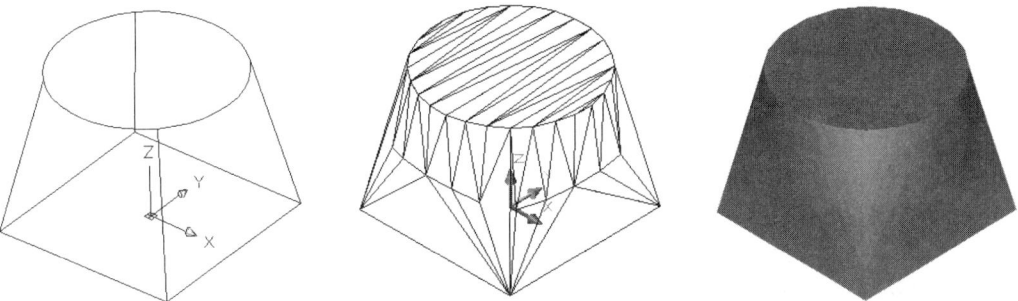

Wireframe, hidden, and fully rendered views of a linear loft

Lofting with the type set to **Cubic** allows you to loft a number of cross section profiles.

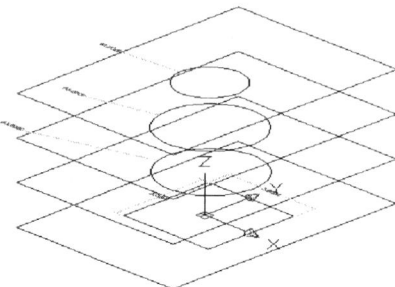

Multiple cross section profiles can be lofted with the Cubic option. Multiple sizes of the same type of profile, or different-shaped profiles, all can be used.

With the Cubic option, the loft is "blended" into each cross section profile. The critical settings for controlling the blending are the weight and angle for each cross section profile.

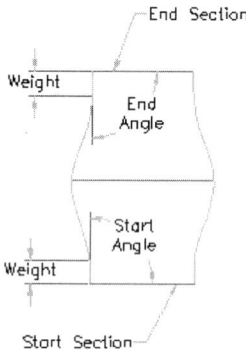

The **weight** value controls how far from the edge of the cross section profile the blend is started. The **angle** is the angle set from the corner of the cross section profile. The angle and weight are set for both a **start** and an **end** section.

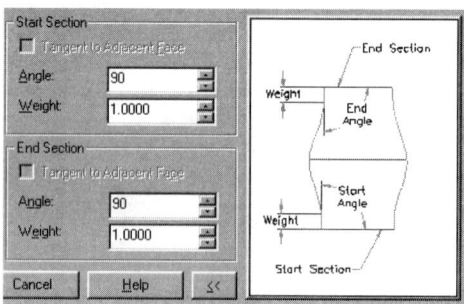

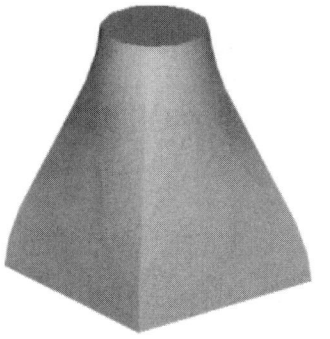

Start
Angle = 90
Weight = 1

End
Angle = 90
Weight = 1

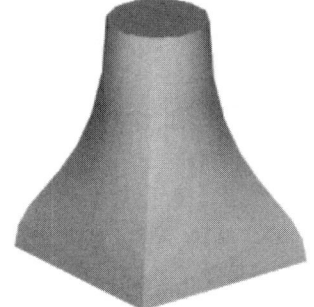

Start
Angle = 90
Weight = 2.5

End
Angle = 90
Weight = 1

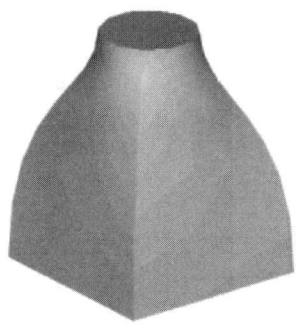

Start
Angle = 90
Weight = 1

End
Angle = 90
Weight = 2.5

Start
Angle = 50
Weight = 2

End
Angle = 90
Weight = 1

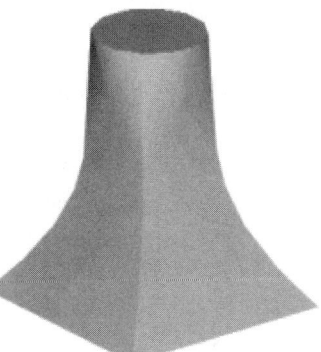

Start
Angle = 90
Weight = 1

End
Angle = 50
Weight = 2

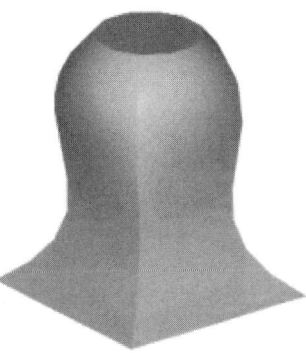

Start
Angle = 140
Weight = 1.5

End
Angle = 35
Weight = 2

The **Closed Cubic** option can be used to create a loft when the first and last cross section profiles are the same. Unique shapes can be created using Closed Cubic. Here we simply used a square and two circles to create the plastic handle.

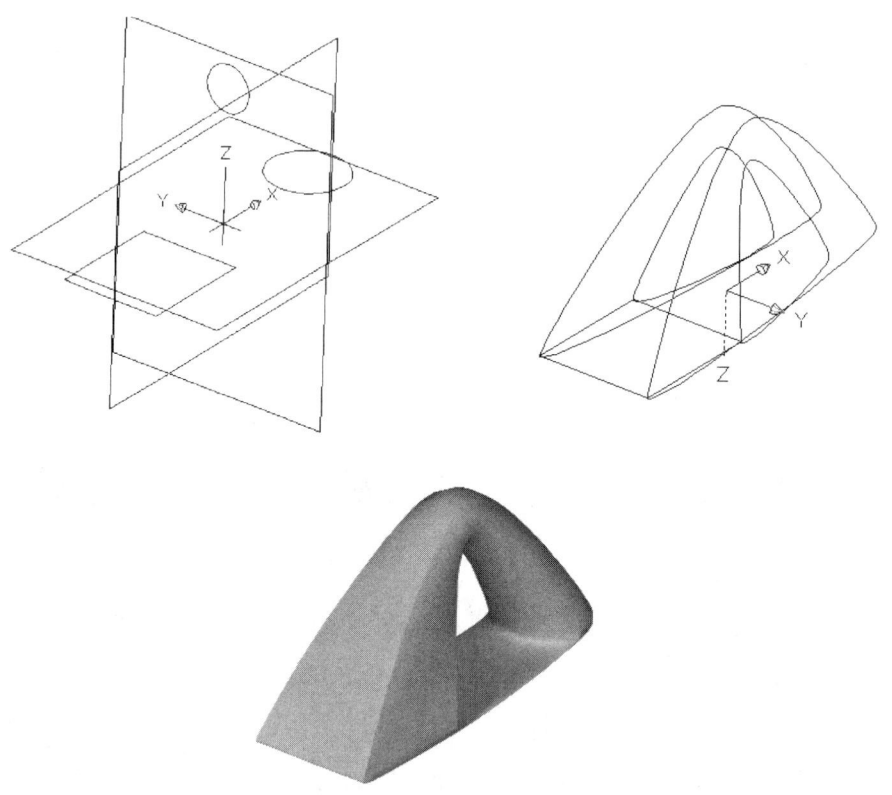

Drawing Project 6–1

Corner Support Bracket

Model the corner support bracket to the exact dimensions indicated. Use Sweep and Extrude as needed, and add the fillets and holes as needed to complete the model. When you have finished, save the drawing and click **8** for an isometric view. Print out a copy for your instructor.

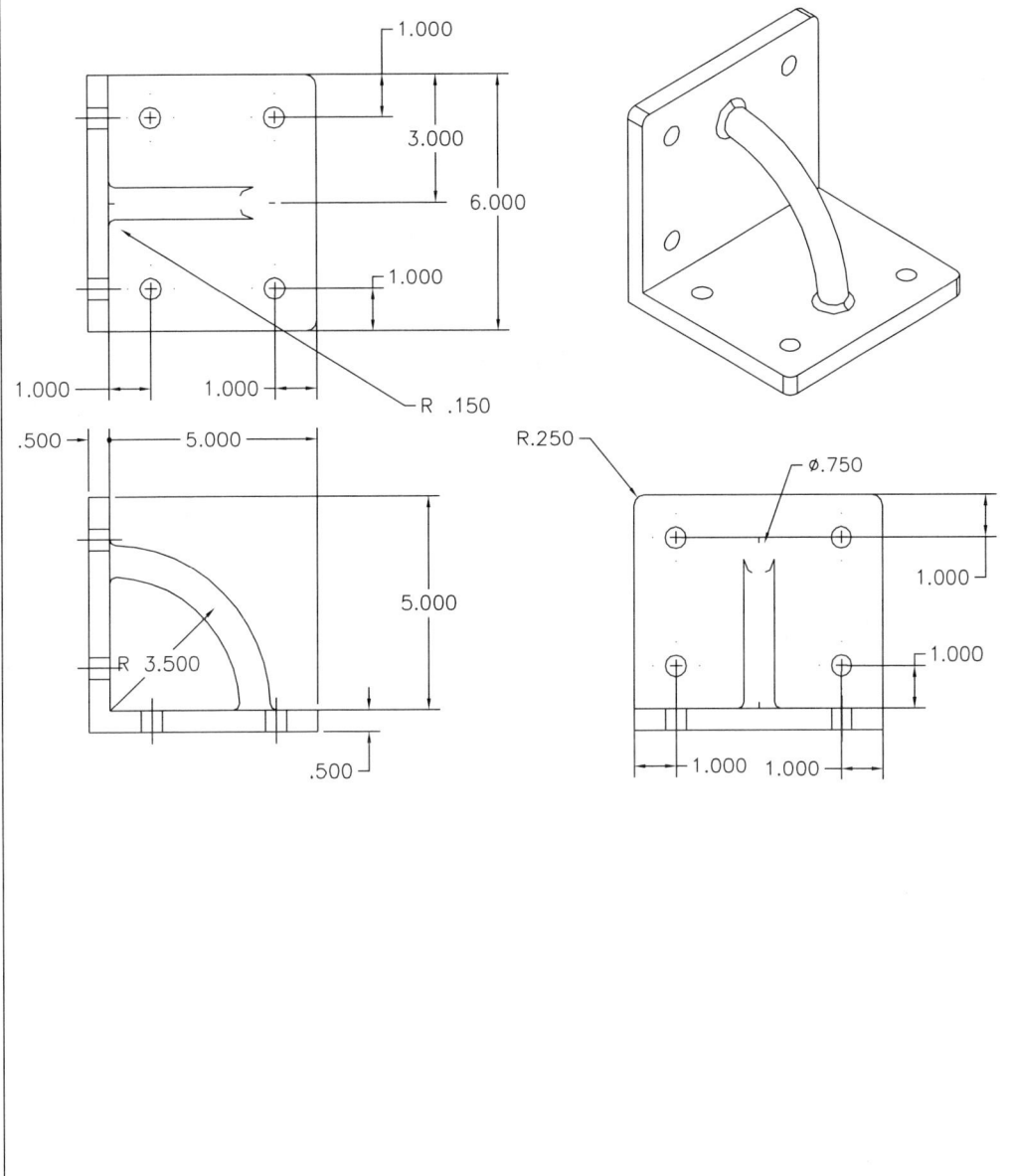

Drawing Project 6–2

Lift Grip Handle

Model the lift grip handle to the exact dimensions indicated. Use Sweep and Extrude as needed, and add the fillets and holes as needed to complete the model. When you have finished, save the drawing and click **8** for an isometric view. Print out a copy for your instructor.

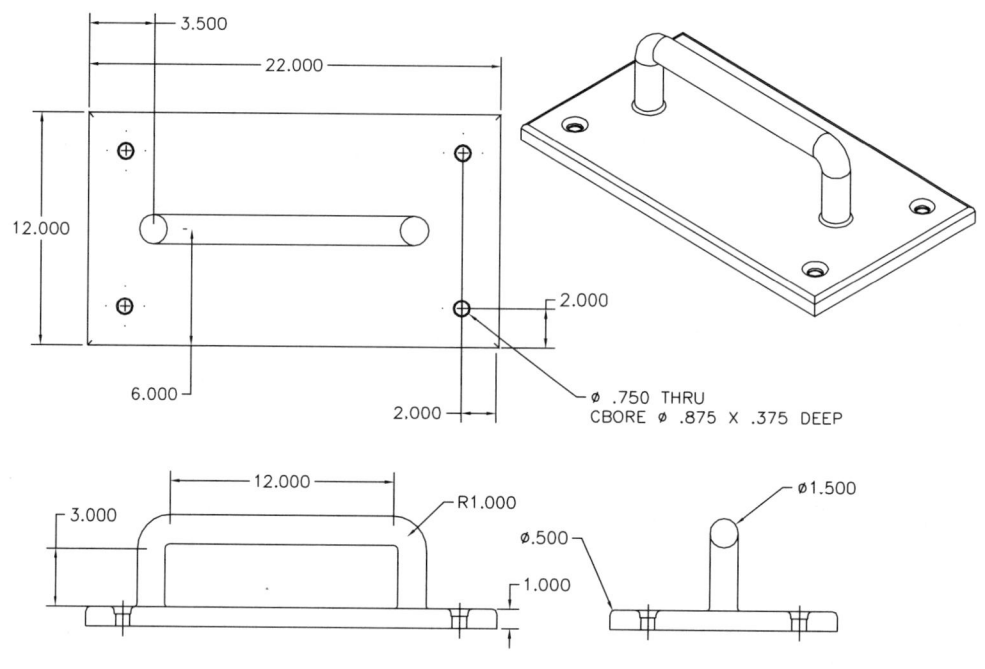

Drawing Project 6–3

Cup

Model the cup to the exact dimensions indicated. Use Sweep and Revolve as needed, and add the fillets. When you have finished, save the drawing and click **8** for an isometric view. Print out a copy for your instructor.

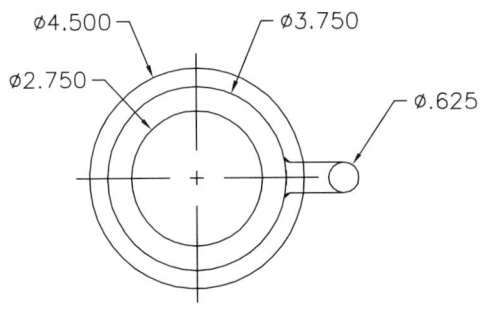
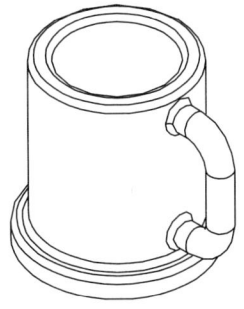
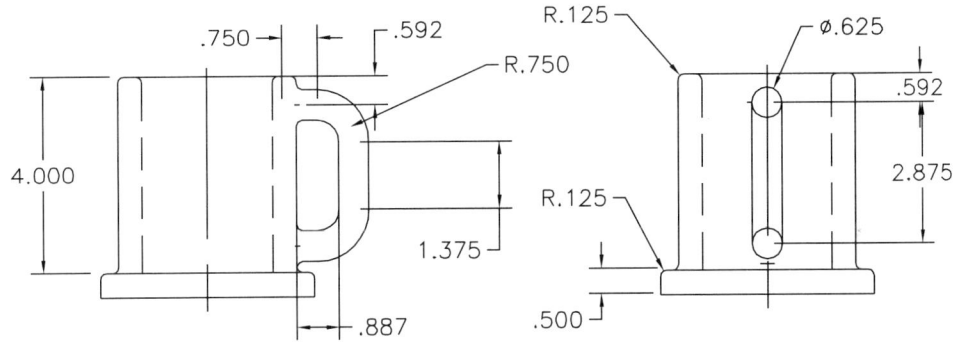

Drawing Project 6-4

Mold

Model the mold to the exact dimensions indicated. Use Sweep and Revolve with the Cut setting. Add the fillets to the mold as needed. When you have finished, save the drawing and click **8** for an isometric view. Print out a copy for your instructor.

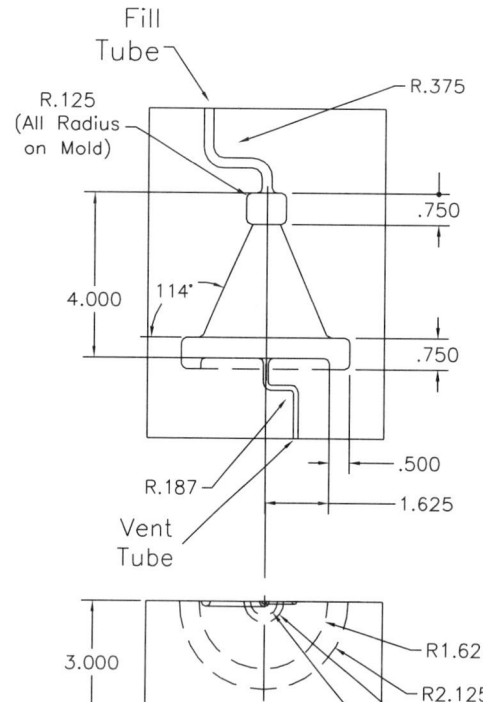

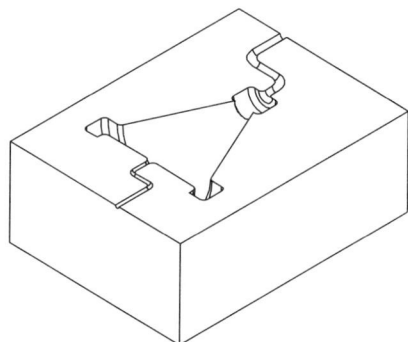

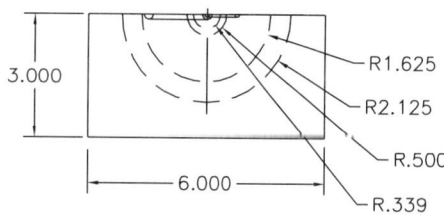

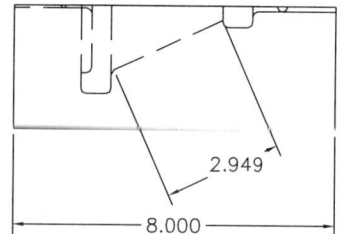

The vent and fill tubes can be any size as long as the diameter and radius is maintained.

Drawing Project 6–5

Stabilizer Arm

Model the stabilizer arm to the exact dimensions indicated. Use Loft and Extrude to create the arm. Add fillets and holes as needed. When you have finished, save the drawing and click **8** for an isometric view. Print out a copy for your instructor.

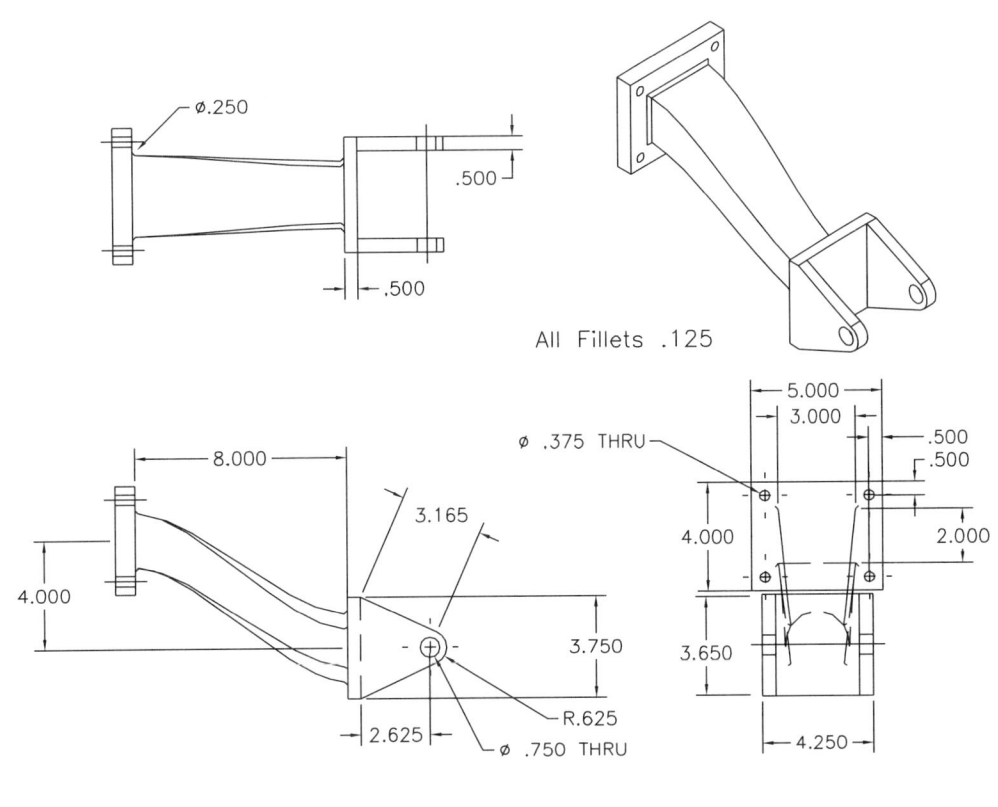

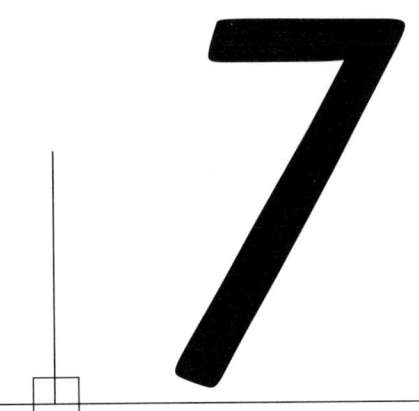

THE SHELL COMMAND

The **Shell** command is a very handy tool for creating many shapes. The shell process "dishes out" the inside of the shape but leaves a "shell" wall thickness around the object. You can then define areas and sides to open to the shell, leaving a refined shape.

Let's create a part using the Shell command.

Step 1   Create the basic profile shape and fully constrain the profile.

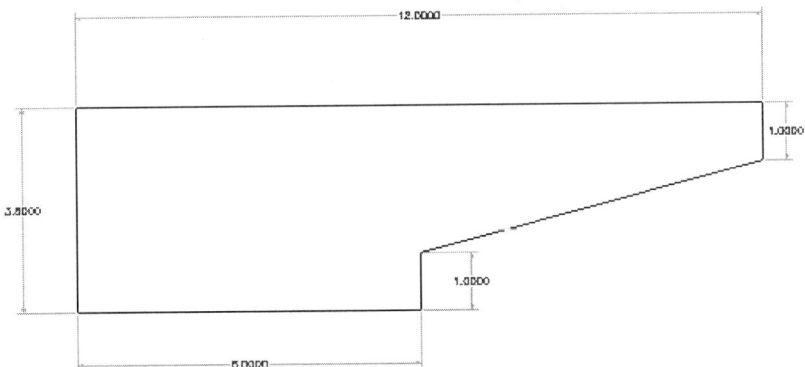

Step 2 Extrude the profile. (Any 3D feature will work here such as Revolve.)

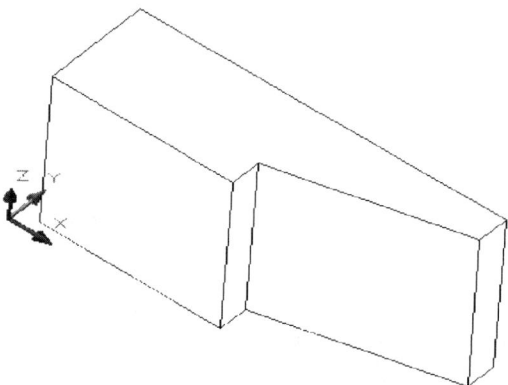

Step 3 Add fillets or any other shape modifications. It is better to add any features that will affect the shape of the final part before executing the Shell command.

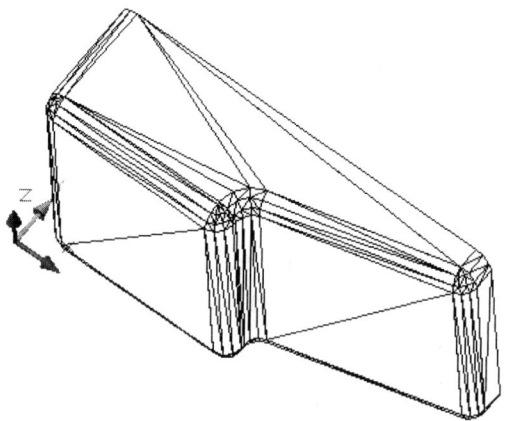

THE SHELL COMMAND | 105

Step 4   Type **AMSHELL** ↵
or click the icon.
The Shell dialog box appears.
You can set the shell to cut the inside, outside, or midplane in the Default Thickness area. If you designed the part to its maximum dimensions, you would shell to the inside. If you designed the part from the interior dimensions, you would add the shell to the outside. You can also adjust the wall thickness in this area.

You usually will need to "open" or exclude the face of one or more sides of the part in the Excluded Faces area. A face can also be reclaimed here.

Click **OK** when you have made the settings.

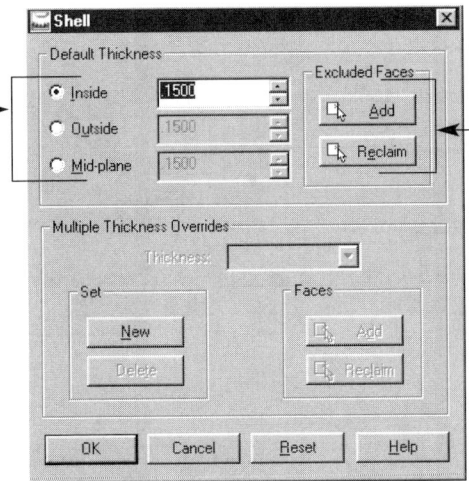

In this example we excluded the top face of the oil pan, set the thickness to .150, and "shelled" to the inside.

For a Deeper Understanding about
Shell

The Inside, Outside, and Mid-plane options set the direction of the shell.

Inside
(Cut to the inside.)

Outside
(Add to the outside.)

Mid-plane
(Both sides of the wall add some thickness, but the total is only the default thickness value.)

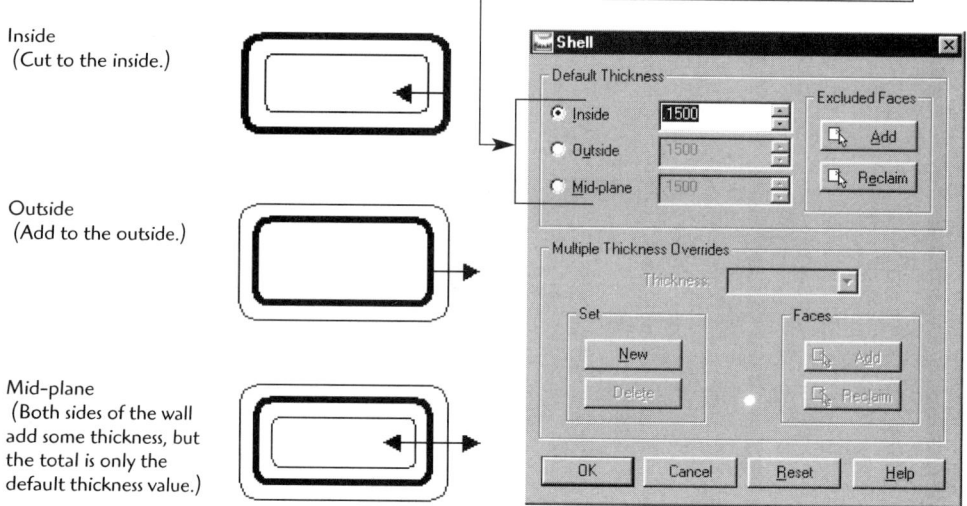

The Exclude Faces area includes the Add and Reclaim options.

This face has been reclaimed.

A face has been added to the faces to be excluded.

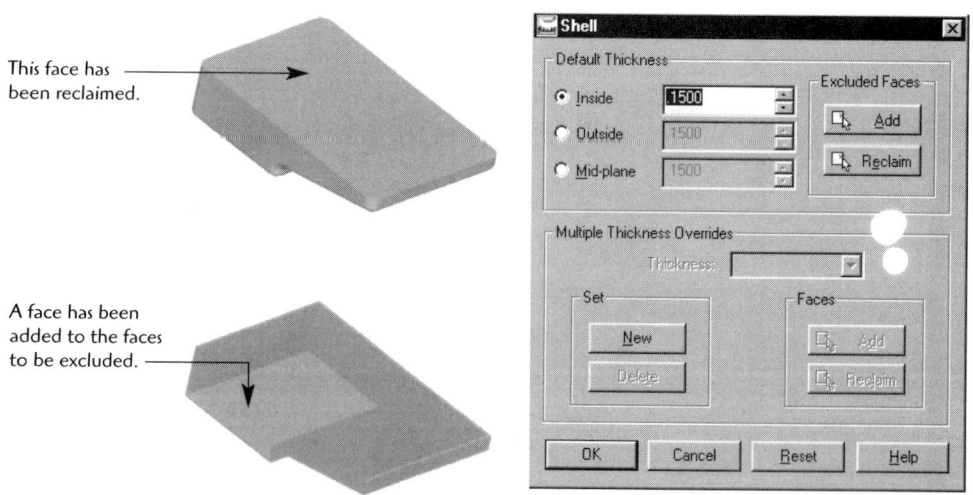

THE SHELL COMMAND | 107

In cases where a wall thickness must be different from the default thickness, you must use the **Multiple Thickness Overrides** options.
To activate these options, click **New** in the Set box area.
Set a new thickness and select Add in the Faces box for the override area.
Click **OK** and update the model.

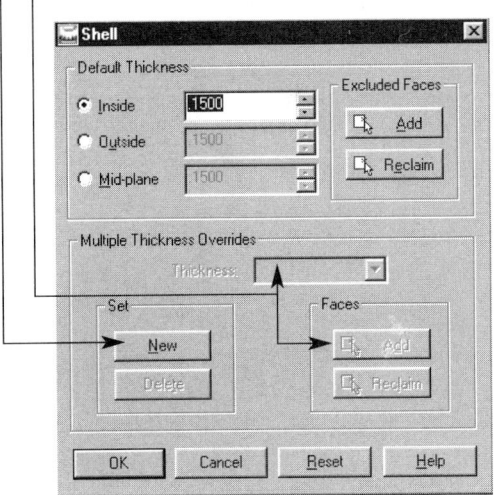

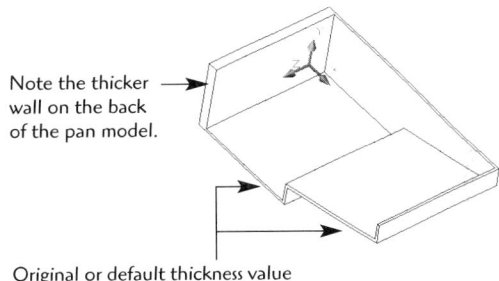

Note the thicker wall on the back of the pan model.

Original or default thickness value

Drawing Project 7–1

Oil Pan

Model the oil pan to the exact dimensions indicated. Use Extrude and Shell to create the oil pan. Add the fillets as needed. When you have finished, save the drawing and click **8** for an isometric view. Print out a copy for your instructor.

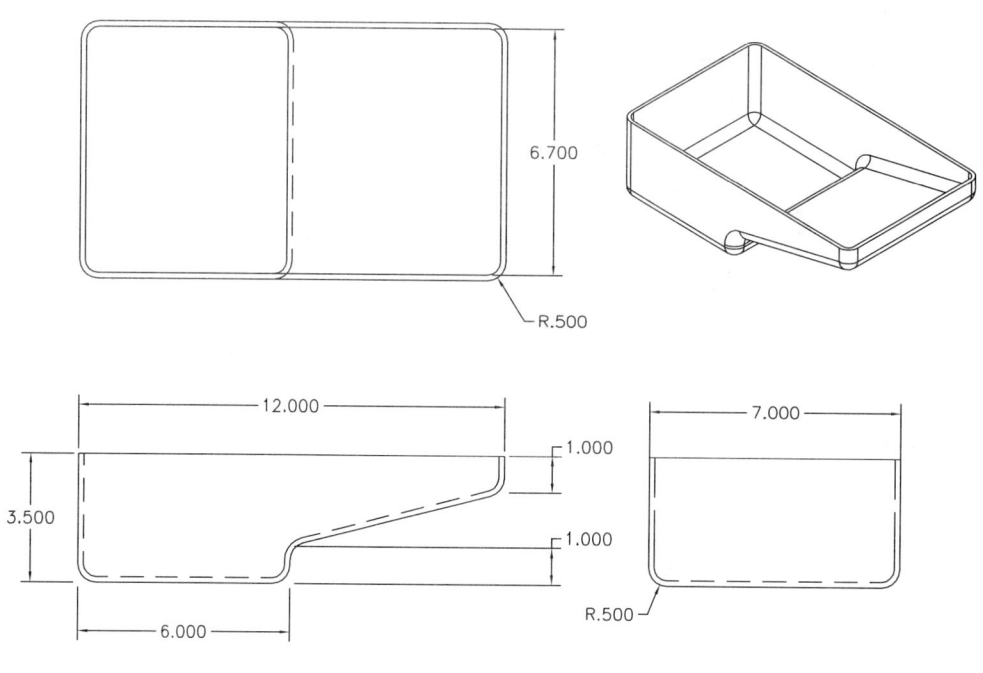

108

CREATING DETAIL DRAWINGS (BLUEPRINTS) FROM PARAMETRIC MODELS

Once you have refined the model to the point it is ready for production, you need to produce detail drawings, or blueprints, for the fabricator. Mechanical Desktop provides excellent detailed drawing capabilities. Standard orthographic view (top, front, and side) details, sections, auxiliaries, isometrics, and assemblies can all be easily produced. Assemblies can be created from completed models, and a **bill of materials** (BOM) also can be added to the drawing. Solid mass property data can be extracted from the model and added to your drawing as well.

In this section we will start with the basic methods of producing detail drawings from your models and progress to the display of assemblies.

The basic steps for creating a detail drawing are as follows:

Step 1   Open one of your previously created models or create a new one.

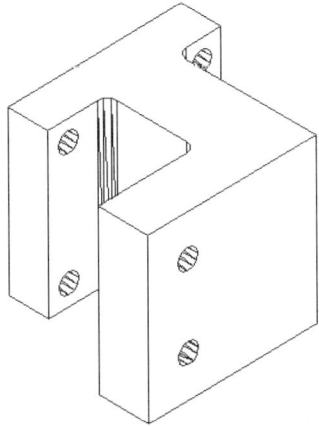

CHAPTER 8

Step 2   Type **AMDWGVIEW** ↵
or click the icon.
The Create Drawing View dialog box appears.
— Set the View Type to **Base**.
— Set the Data Set to **Active Part**.
— The **Scale** is set low but can be increased later. (Use .375.)
Leave other settings on the defaults.
Click **OK**.

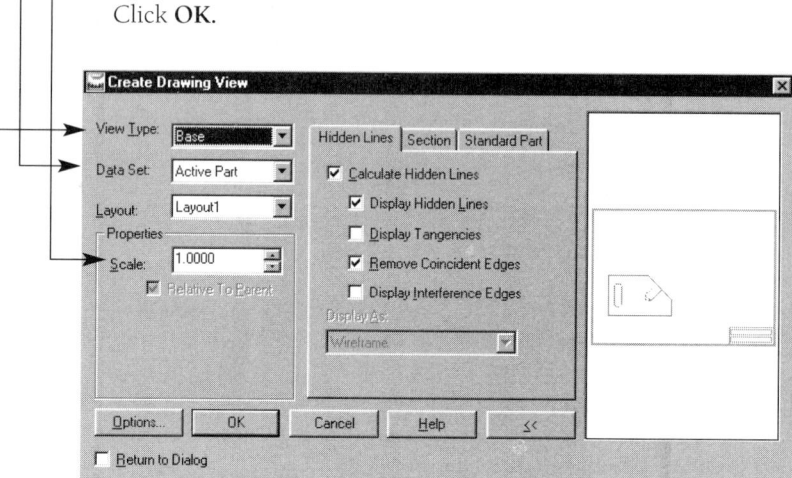

Step 3   You must now tell Mechanical Desktop how you want this base view to be aligned and displayed. Once you pick the face you want displayed, you must indicate how it is to be positioned.

— **Select** the face you want to be the base ↵. (This is usually referred to as the front view.)

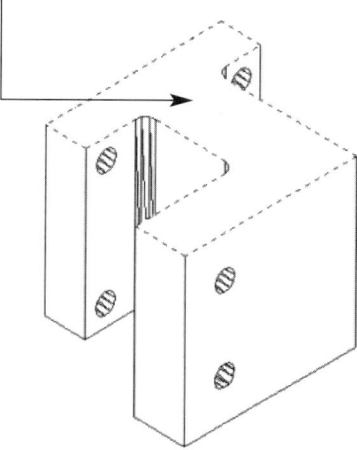

Type **Y** ↵. (*Note:* This gives you a Y-axis starting point.)
Type **R** ↵.
(*Note:* Continue to type R until the XYZ axis icon is aligned in the standard Cartesian coordinate layout that will represent the front view.)

- X is left/right.
- Y is up/down.
- Z is in/out.

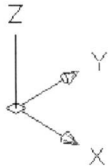

Step 4   Specify the location of the base view.
(*Note:* Pick where you want the view to be located. Pick the lower left of the Layout 1 screen. This is the typical location for the **front view.**)

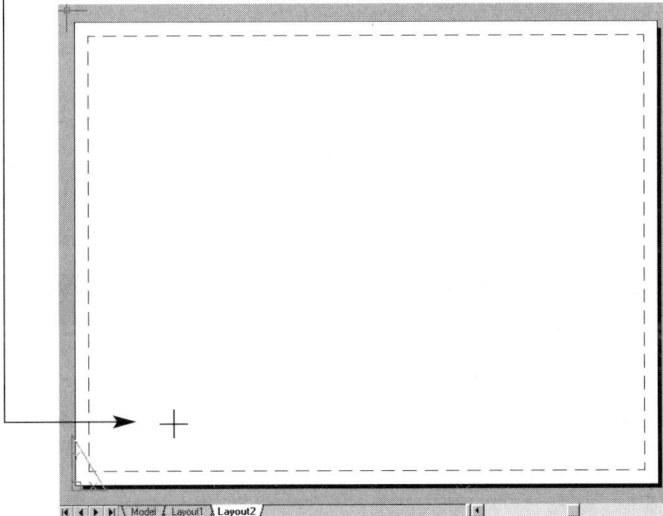

112 | CHAPTER 8

— The base view is in place.

Step 5 You now need to repeat the view creation process for the other views.
Type **AMDWGVIEW** ↲
or click the Drawing View icon.

The Create Drawing View dialog box appears.
Set the View Type to **Ortho**.
Leave other settings on the defaults.
— Click **OK**.

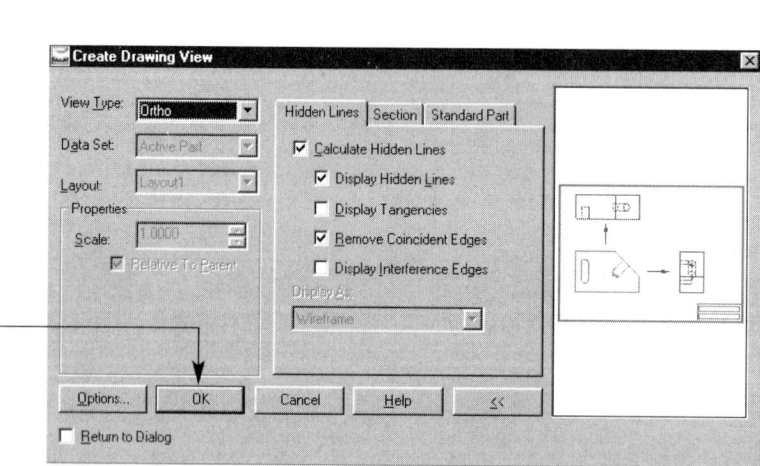

CREATING DETAIL DRAWINGS | 113

Step 6 Select the parent view.

(*Note:* You must select the parent view, which is the base view just created.)

Drag the line up to the position for the new view, click and ↵.

The new view is placed.

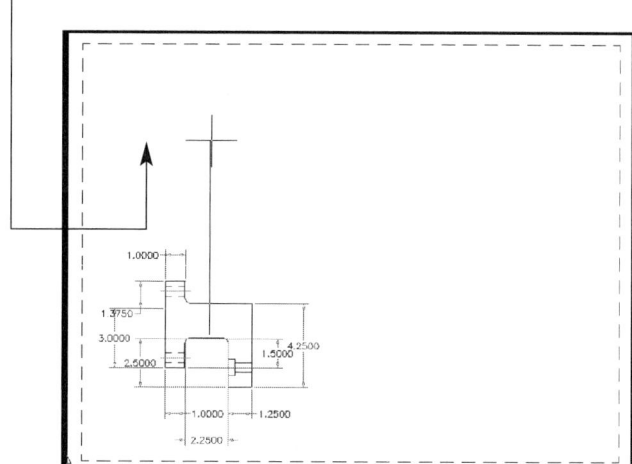

Repeat steps 5 and 6 for the next (side) view.

The next step is to create an isometric (iso) view. The step is the same as for the other views, except that Iso is selected as the view type.

Type **AMDWGVIEW** ↵
or click the Drawing View icon.
The Create Drawing View dialog box appears.

In the View Type select **Iso**.
The Scale is relative to the other views, so leave it set to 1.0000.
Click **OK**.

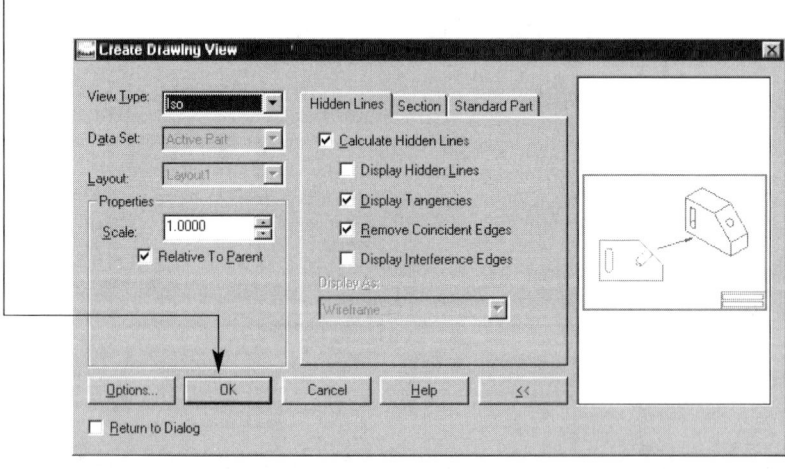

Select the parent view again and **drag** it into the remaining corner.

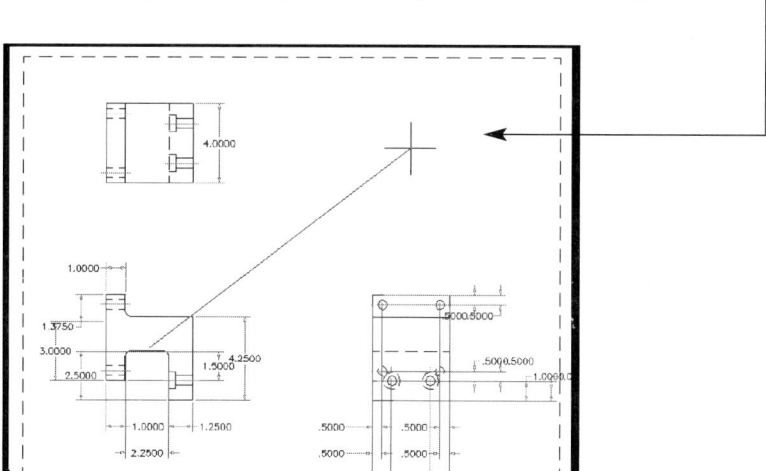

The placement of the views is complete.

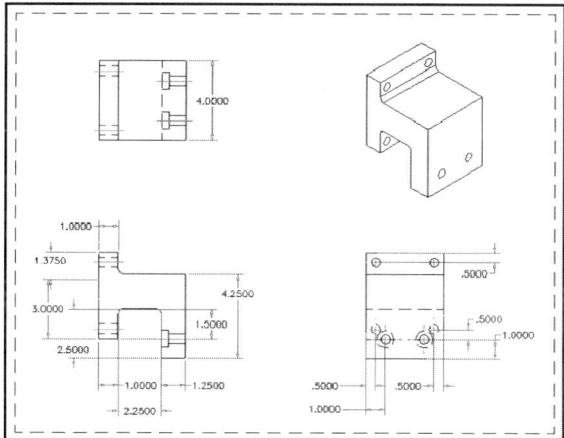

CHAPTER 8

Much more can be done to detail and enhance the drawing, as you will see in the Deeper Understanding section. A helpful toolbar to help you keep track of some of the drawing layout options is the Drawing Layout toolbar:

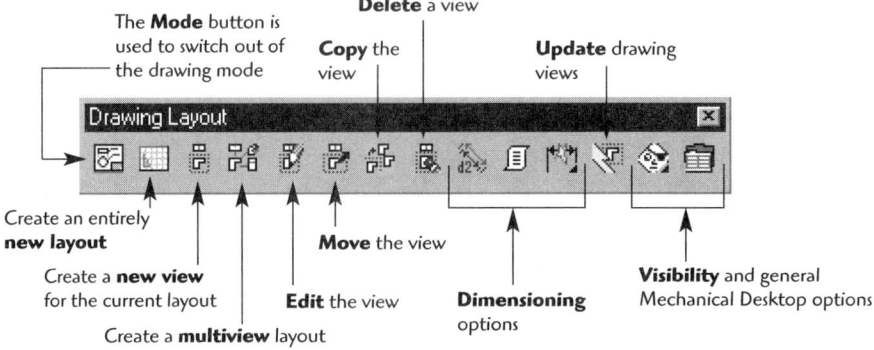

The **Mode** button is used to switch out of the drawing mode
Copy the view
Delete a view
Update drawing views
Create an entirely **new layout**
Move the view
Visibility and general Mechanical Desktop options
Create a **new view** for the current layout
Edit the view
Dimensioning options
Create a **multiview** layout

For a Deeper Understanding about

Detail Drawings (Blueprints)

Let's first look at how you would edit the views you have created.
Type AMEDITVIEW ↵
or click the **Edit View** icon.

Select view to edit: *(Click on one of the views you have created.)*

The Edit View dialog box appears.

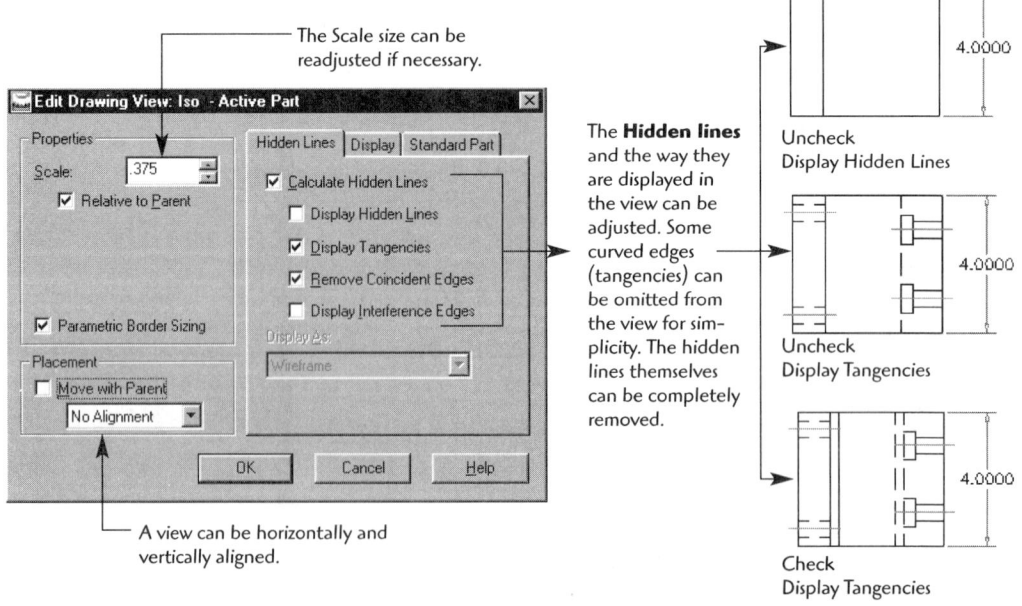

The Scale size can be readjusted if necessary.

A view can be horizontally and vertically aligned.

The **Hidden lines** and the way they are displayed in the view can be adjusted. Some curved edges (tangencies) can be omitted from the view for simplicity. The hidden lines themselves can be completely removed.

Uncheck Display Hidden Lines

Uncheck Display Tangencies

Check Display Tangencies

CREATING DETAIL DRAWINGS | 117

The **Display** tab options are as follows:

Holes that are tapped (threaded) have special hidden lines to indicate the cut threads. This display can be turned on and off.

The **constraining dimensions** can be turned on or off in the view.

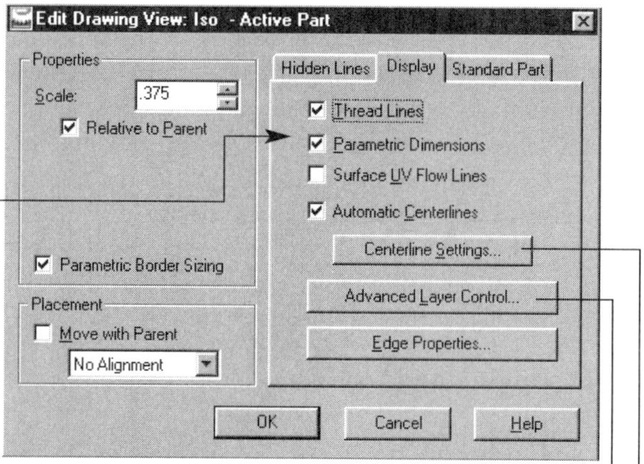

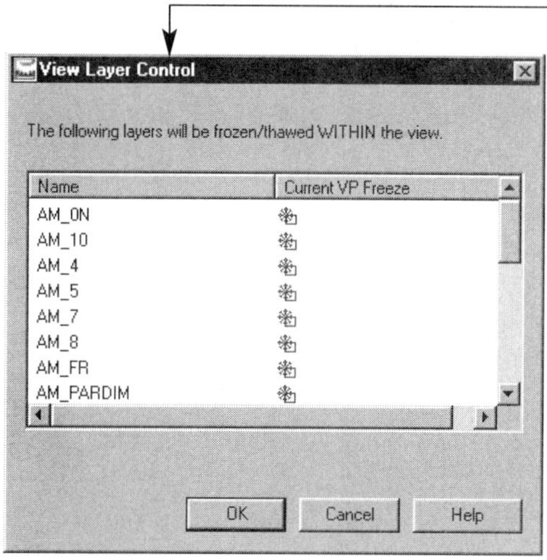

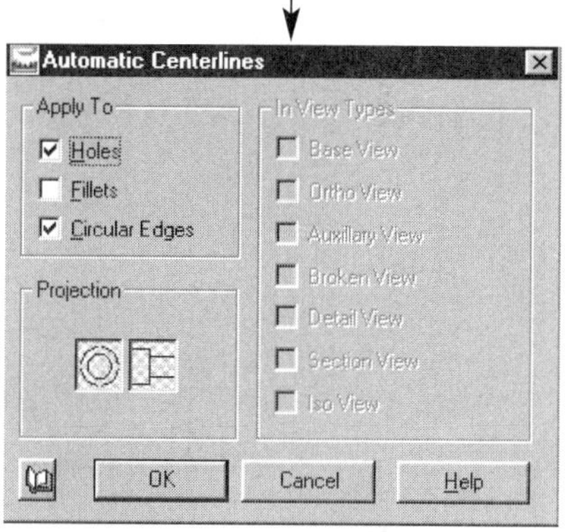

Centerlines (Long-short-long) can be applied to holes, fillets, and circular edges in a view. These lines form an extension line for necessary dimensions.

Viewport layers are layers unique to each viewport. The display visibility of viewport layers can be adjusted.

How about moving the views?

Type AMMOVEVIEW ↵
or click the icon.

Specify view to move: *(Click on the view, then drag the view to the new location.)*

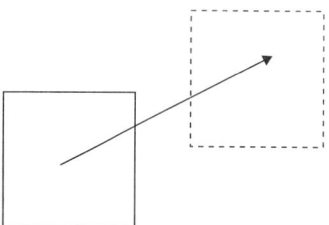

When using the AMMOVEVIEW command (move view) you may note that the Base and Iso views will move in all directions, but the views created as Orthos will move only on one axis. This is because of the orthographic constraints placed on the view. A top view will move up and down, but not left and right. A side view will move left and right but not up and down due to the orthographic constraint. Allowing a view to move out of these limitations would mean it was *not* an orthographic projection. If you need to make a view movable in all directions and out of the orthographic alignment, create it as another **base** view. You will have to define and align the view like any other base view, but now you can move it anywhere on the drawing layout. It is better to keep views orthographic to the original base view and let Mechanical Desktop keep them properly aligned in the drawing and readable for technicians and engineers.

How about copying a view?

Type `AMCOPYVIEW` ↵

or click the icon.

`Specify view to copy:` *(Click on the view, then drag the copy to the new location.)*

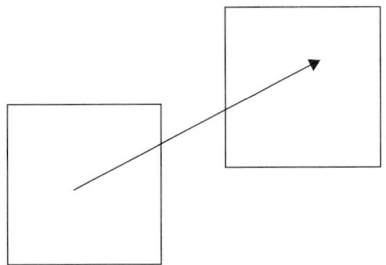

How about deleting a view?

Type `AMDELVIEW` ↵

or click the icon.

`Select view to delete:` *(Click on the view to delete it.)*

Hint: Remember that the UNDO command or the Undo icon is available if you need to "back up" due to a mistake.

What if you need to know exactly what settings are applied to a view? In other words, are the tangents or hidden lines turned on? You may need this information to keep all the views consistent or to display special details of a particular view. Use the AMLISTVIEW command to list this information.

CREATING DETAIL DRAWINGS | 119

Type **AMLISTVIEW** ↵
or click the icon.

Select the view: *(Click the view about which you need the information.)*

Base Drawing View
id = 1
Name = Base   view is ACTIVE and up to date

       View scale: .3750
  View direction: .0000,.0000,1.0000
    Center point: 2.4761,2.4058
    Target point: .0000,.0000,.0000

Hidden lines are displayed.
Tangent edges are not displayed.
Automatic viewport sizing is enabled.
View has 3 descendants, 9 dimensions, 3 notes.
View is not aligned to any other view.
View has 0 views aligned to it.
One part represented.

All view setting information is listed in the text box.

Click the X in the upper right to close the text box.
Once you examine the list you can go back and change settings with the AMVIEW command. Follow the procedure previously discussed for AMVIEW.

Setting other drawing display options

Type **AMOPTIONS** ↵.

Borders, edge colors, and other options can be adjusted in this box.
Click **OK** when you have completed your setting adjustments.

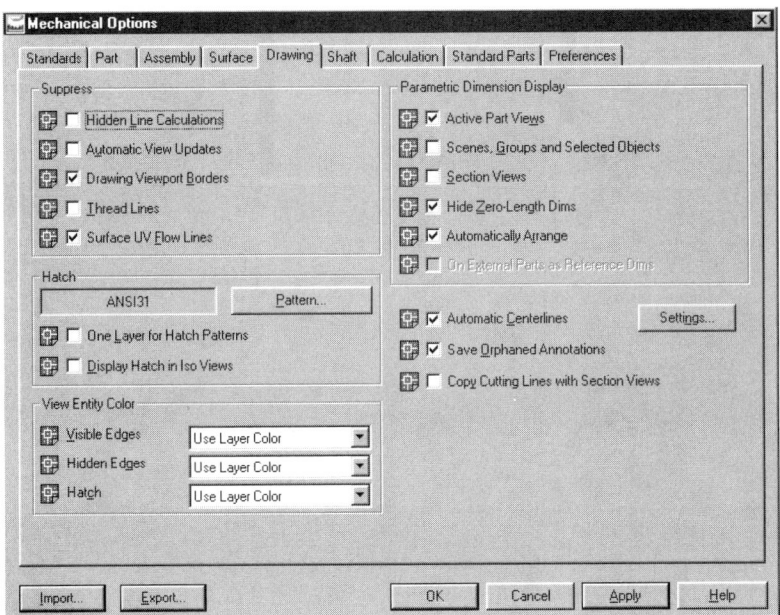

How about creating an entirely new, additional drawing layout?

You may need to represent the part or component for different customers, contractors, or fabrication groups. Mechanical Desktop provides an easy method for creating additional drawing layouts.

Type **Layout** ↵.
Type **New** ↵.
Type in a unique layout name, then ↵.

You will now see the new layout tab at the bottom of the drawing page. Clicking on this tab opens the new blank layout page, and you can add views with the same procedures that you have used previously.

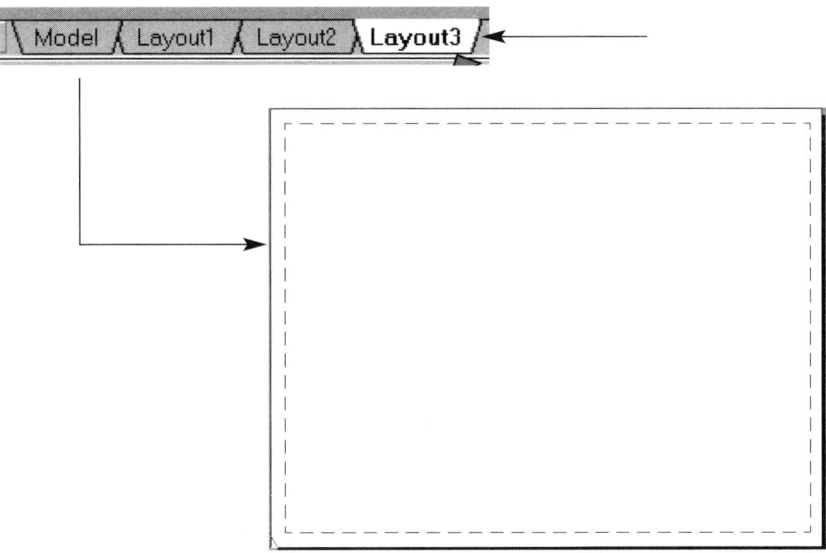

Type **AMDWGVIEW** ↵.
Adjust settings as discussed earlier and click on the screen to place the view.

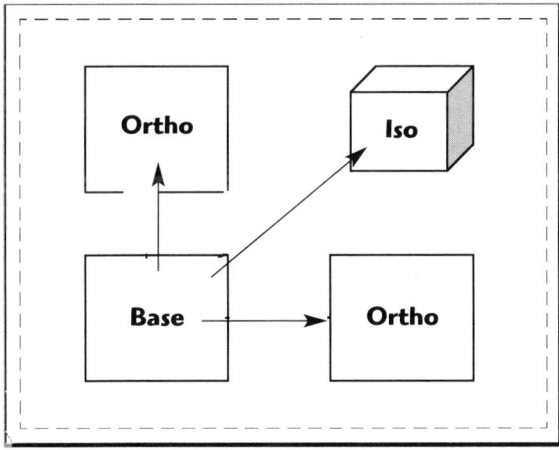

ADDING AND ADJUSTING DIMENSIONS ON A DRAWING

When you created each view on your detail drawing, you noted that any dimensional constraints that were created in the model mode came into the drawing view as standard orthographic style dimensions. These dimensions are usually helpful and necessary for the drawing, but additional dimensions are usually necessary. Mechanical Desktop has handy intuitive commands for adding these dimensions. For example, fillet, radiis, holes, and the like all need to be added to the drawing.

Type **AMREFDIM** ↵.

`Select first object:` *(Click on the object area to be dimensioned such as a fillet or line. In this example we clicked on a radius.)*

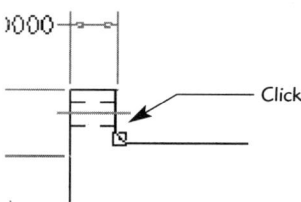

`Select second subject or place dimension:` *(Click where you want the dimension to be located.)*

The **AMREFDIM** (reference dimension) is placed in the view.

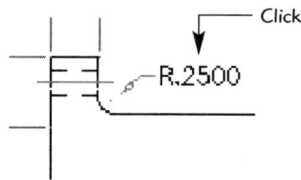

How about dimensioning holes?

You may use the AMREFDIM command, but a better choice is the AMNOTE command (AMHOLENOTE on older versions of MDT). The AMNOTE command adds all the required hole information to the leader line.

Type **AMNOTE** ↵
or click the icon.

`Enter an option [New]:`↵ *(Press Return to accept the New option.)*

`Select the hole feature:` *(Click on the hole in the view that best represents the hole.)*

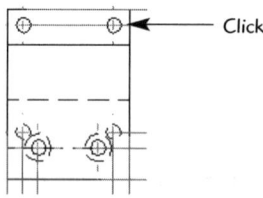

CREATING DETAIL DRAWINGS | **123**

The Hole box appears, where you can adjust settings, as well as the amount of information to be included in the note.

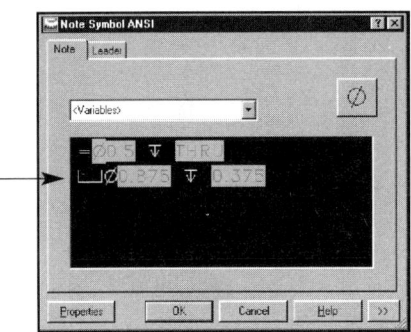

Usually if the default ANSI (American National Standards Institute) setting style is applied, your hole notes will contain the correct symbol information.

Click **OK**.

`Select location for hole note:` *(Click the location for the note.)*

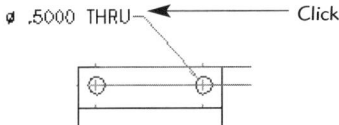

Let's add another hole, this time a counterbored hole. (Use the same AMNOTE procedure for the countebored hole as for standard holes.)

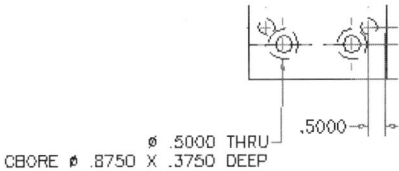

To **edit** a hole note, **double-click** on the dimension, or type **AMPOWEREDIT** ↵.
The editor box appears, where you can type in changes, additions, or deletions.

Click **OK** when done.

You can add additional symbols by clicking on the symbol button.

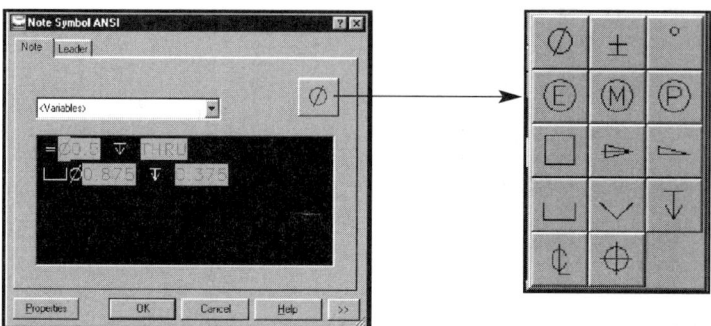

Moving Dimensions

Grips can be used to further move the dimensions. Grips are the blue boxes that appear when an object is selected when the command line is blank.

Command: *(There is no command active, so the command line is blank.)*

Click on the object: *(In this case click on the dimension object, which turns red.)*

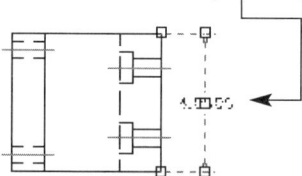

Drag the red box to the new location and click the Esc key two times to clear the grip boxes.

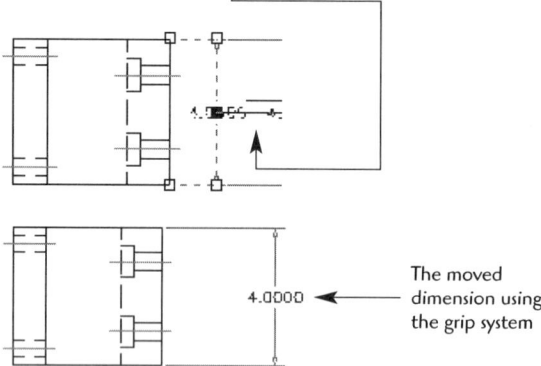

The moved dimension using the grip system

Aligning dimensions

The figure below shows the current dimension positions in the orthographic drawing.

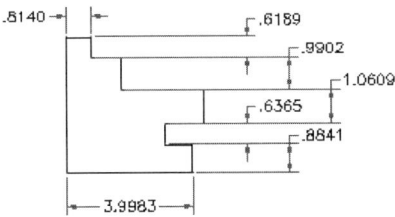

Type **AMDIMALIGN** ↵.

Select base dimension: (*Select the dimension to which all other dimensions are to be aligned.*)

Select linear dimensions to align: 1 found (*Now, click all other dimensions that need aligning. The first selection is confirmed.*)

Select linear dimensions to align: 1 found, 2 total (*Mechanical Desktop confirms the second selection.*)

Select linear dimensions to align: 1 found, 3 total (*Mechanical Desktop confirms the third selection.*)

Select linear dimensions to align: 1 found, 4 total (*Mechanical Desktop confirms the fourth selection.*)

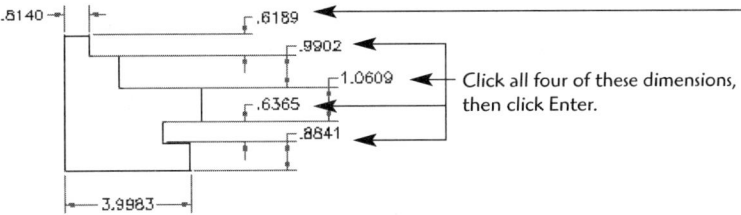

Press ↵.
(*Note:* Pressing the Enter key always confirms you are done selecting.)

4 dimension(s) aligned. (*Mechanical Desktop responds that four dimensions were aligned.*)

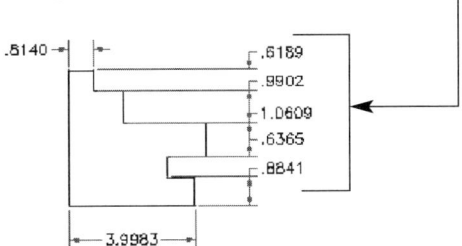

Inserting a Dimension (in-line with an existing dimension)

Here a dimension exists, but details have been added and additional dimensions need to be added, and we would like to place them in-line with the existing dimension line.

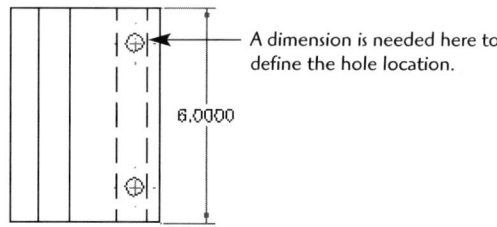
A dimension is needed here to define the hole location.

Type AMDIMINSERT ↵.

Select base dimension: *(Click the original or existing dimension.)*

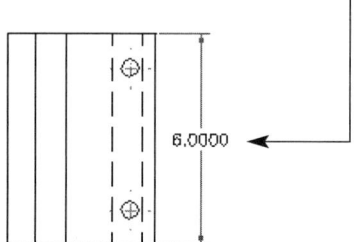

Locate extension line origin: *(This is the new extension line to be inserted. Use an osnap to accurately place the extension line. Here CEN was used for the center osnap to the hole.)*

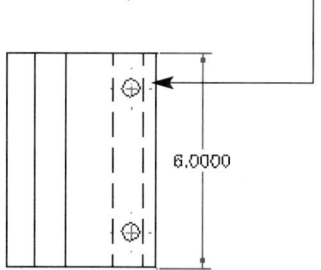

Note that the inserted dimension has been accurately placed directly in-line with the original dimension line.

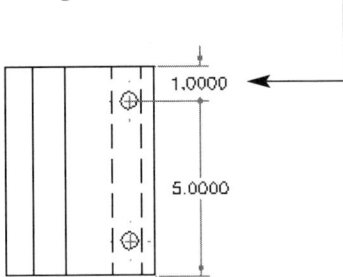

The process can be reversed by using **AMDIMJOIN**, the join dimension command. Type `AMDIMJOIN` ↵.

`Select base dimension:` *(Click the original dimension.)*

`Select linear dimensions to join: 1 found` *(Click the dimension to be joined to the original.)*

`Press` ↵. *(Pressing the Enter key confirms you are done selecting.)*

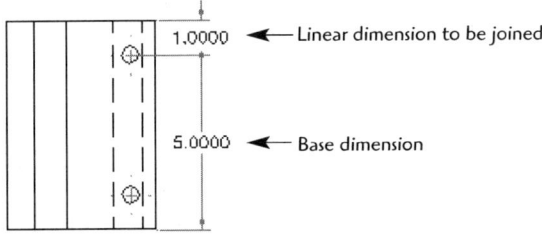

`1 dimension(s) joined.` *(Mechanical Desktop responds that the two dimensions were joined into one.)*

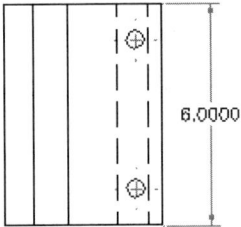

Sometimes the extension lines on crowded dimensions can overlap, causing confusion for the reader. The **AMDIMBREAK** command opens up an area through which the intersection lines can pass. The "broken" lines will maintain their associative properties with respect to editing.

Let's try an example of the AMDIMBREAK command.

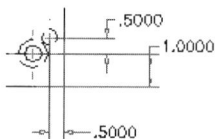

A confusing array of extension lines

Type **AMDIMBREAK** ↵.

`Select dimension or extension line to break <Multiple>:` *(Click the line to be broken.)*

`Second point or [First point/Objects/Restore] <Automatic>:` *(Click the line that crosses the line to be broken or accept the "automatic" option by pressing Enter.)*

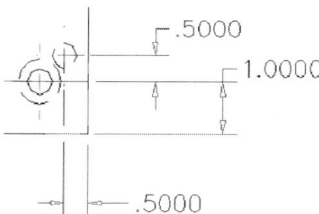

Line breaks applied clarifying extension
line points of the dimension

ADVANCED ORTHOGRAPHIC DRAWING PROJECTIONS

Front, top, and right-side orthographic drawings make up the basic views for detail drawing projections. Views such as **sections, details,** and **auxiliary views** can clarify and further define the intent of the designer or draftsperson. In this next section we will look at the steps involved in applying these detailed views to a model.

CREATING DETAIL DRAWINGS | 129

Let's look at the steps for creating *section views*. Usually, you start with a completed model and the basic orthographic layouts completed and a work plane created at the point where you require the section view.

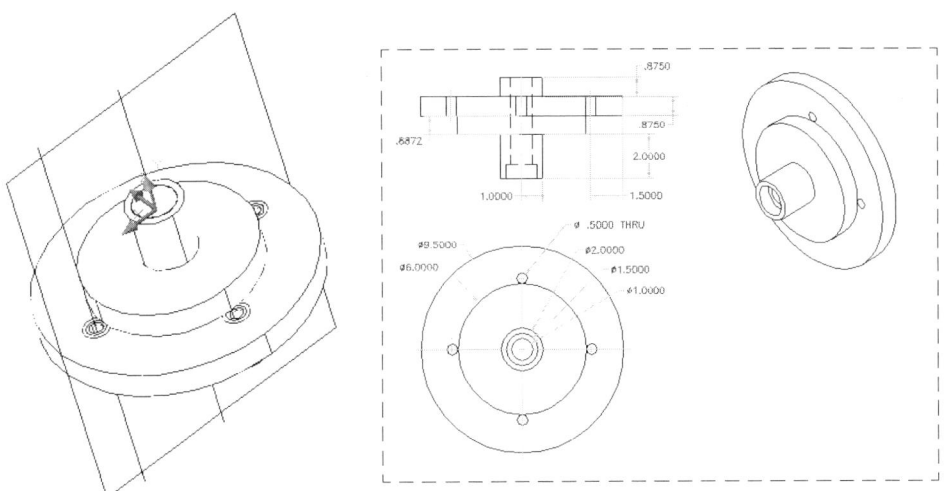

The basic three views are created in the drawing mode, and a work plane is placed where the section is to be cut.

Type **AMDWGVIEW** ↵
or click the icon.

The Create Drawing View dialog box appears.
— Set the View Type to **Base** or **Ortho**.
— Set the Data Set to **Active Part**.
— Set an appropriate **Scale**.
Click on the **Section** tab, and set the Type to **Full** (for full section view).

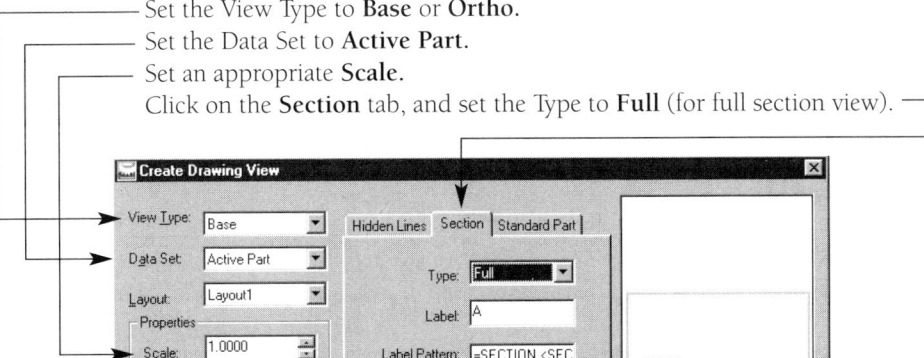

Click the Pattern... button.
The Hatch Pattern dialog box appears.

Leave the default **ANSI31 pattern,** or change to the desired pattern.

The hatch **Scale** can be adjusted. Larger numbers produce a larger space between hatch lines; smaller numbers produce a tighter space between hatch lines.

The **Angle** is usually set or left at 0 but can be changed.

The **Exploded** hatch option is usually not checked unless 2D editing such as trimming of the hatch lines is required.

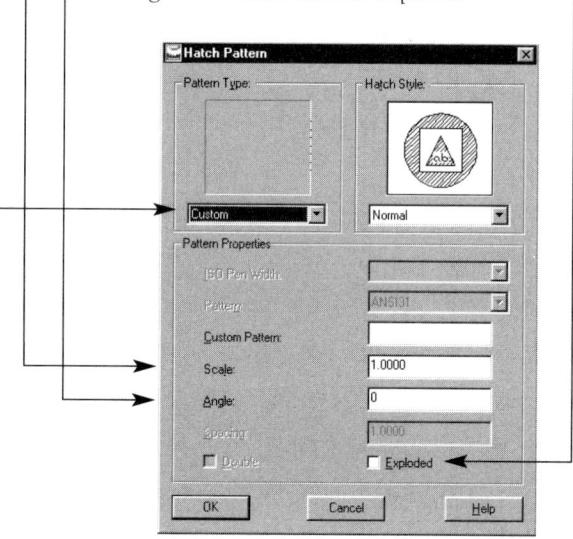

Click **OK**.
Leave the remaining setting on the defaults.

`Selected plane will be the cutting plane.` *(Click on the work plane in the model mode. Mechanical Desktop will automatically switch into the model mode to make the work plane selection.)*

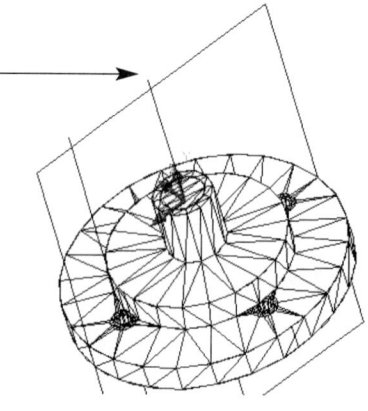

CREATING DETAIL DRAWINGS | 131

Select work plane, face or [worldXy/worldYz/worldZx/Ucs/View]:Y
(*Type* **y** *or* **x**, *and then press Enter.*)

Enter an option [Rotate/Flip/Accept] R (*Type* **R** *for rotate. Rotate until the X, Y, and Z icon is correctly aligned for the section view in the drawing mode.*)

Rotate until:
X is left and right
Y is up and down
Z is forward and back

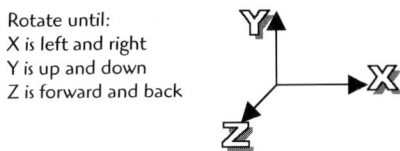

Specify location on base view: (*Click on the drawing mode screen where the section view should be placed.*)

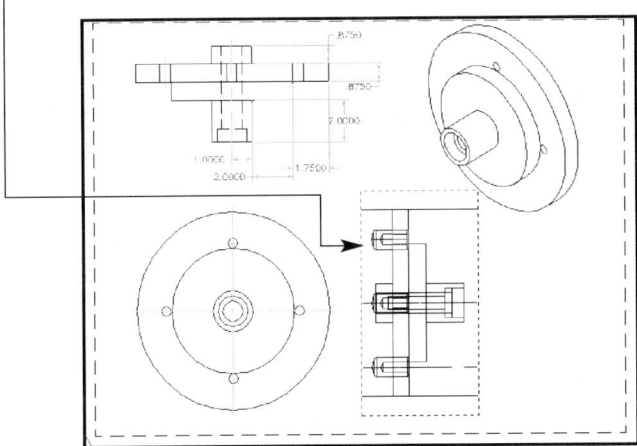

Select view in which to display cutting lines (or press Enter for none): (*Click into a view to place the long-short-short-long cutting plane line. Or type* **P** *for point, and click on a point to display the cutting plane lines. If you press Enter, no cutting plane line will be placed.*)

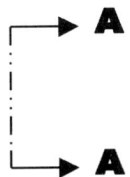

CHAPTER 8

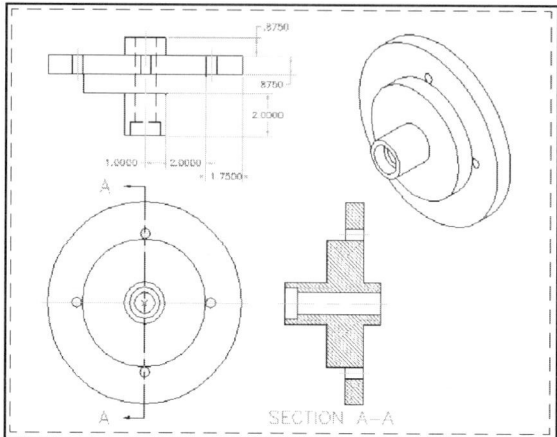

The completed "Section A-A" full section view incorporated into the existing orthographic drawing

Now, let's look at the steps for creating *detail views*. Again, start with a completed model and the basic orthographic layouts completed.

Type **AMDWGVIEW** ↵
or click the icon.

The Create Drawing View box appears.
— Set the View Type to **Detail**.
— Adjust the **Scale** to enlarge the view slightly.
Click **OK**.

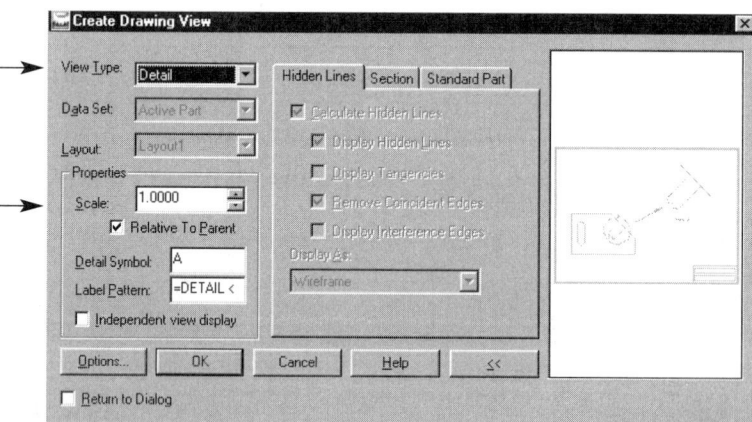

CREATING DETAIL DRAWINGS | 133

Select vertex in parent view to attach detail: *(Click directly in the center of the detail area of the parent view.)*

Specify center point for circular area or [Ellipse/Polygon/Rect/Select]: *(The shape of the detail area can be adjusted to an ellipse, polygon, or rectangle. For the default circle, click directly in the center of the detail area of the parent view.)*

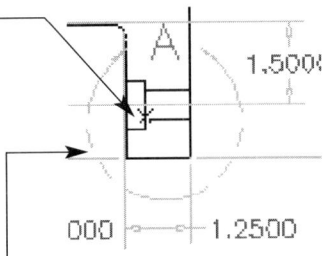

Specify radius of circle or [Diameter]: *(Drag the diameter to cover the area to detail.)*

Specify location for detail view: *(Click in a clear area to place the detail view.)*

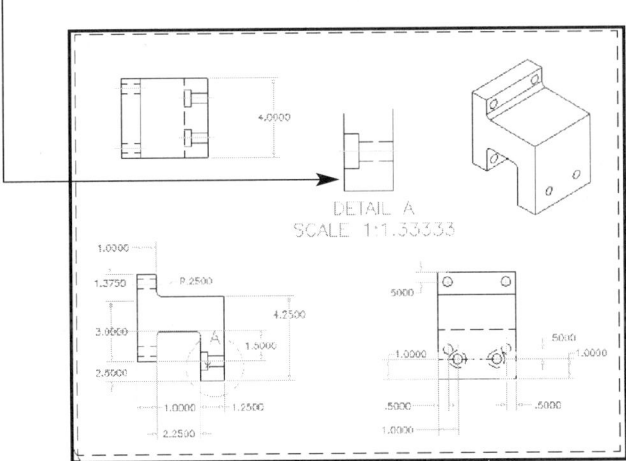

The completed Detail - A view incorporated into the existing orthographic drawing

Auxiliary views are used on slanted surfaces that would not provide a clear view in a typical orthographic projection.

Now, let's look at the steps for creating auxiliary views. Start with a completed model and the basic orthographic layouts completed.

Type **AMDWGVIEW** ↵
or click the icon.

The Create Drawing View dialog box appears.
Set the View Type to **Auxiliary**.
Click **OK**.

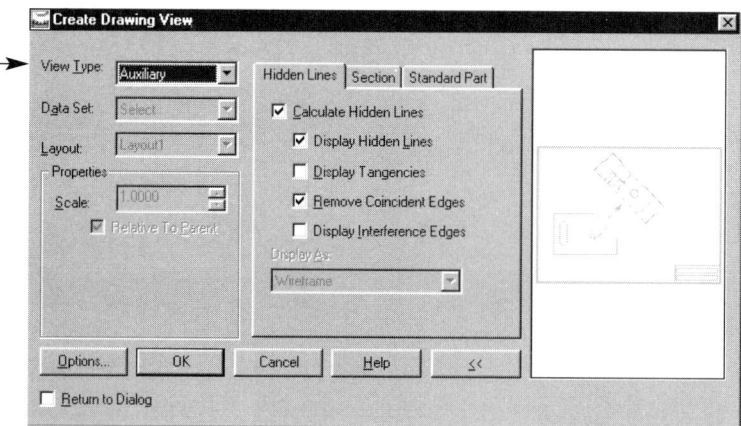

```
Select first point for projection direction or [Workplane]:
```
(*Click a point on the corner of the angled surface. Note that a work plane can also be used to define the surface.*)

```
Select second point or <ENTER> to use the selected edge:
```
(*Click another point on the corner of the angled surface.*)

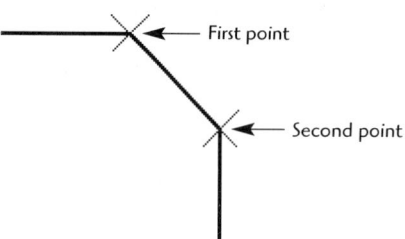

```
Specify location for view:
```
(*Drag into a clear area to place the auxiliary.*)

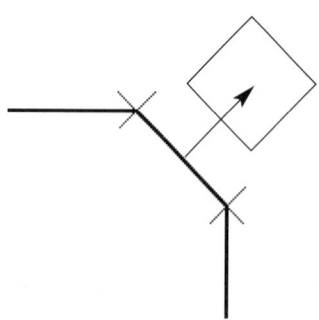

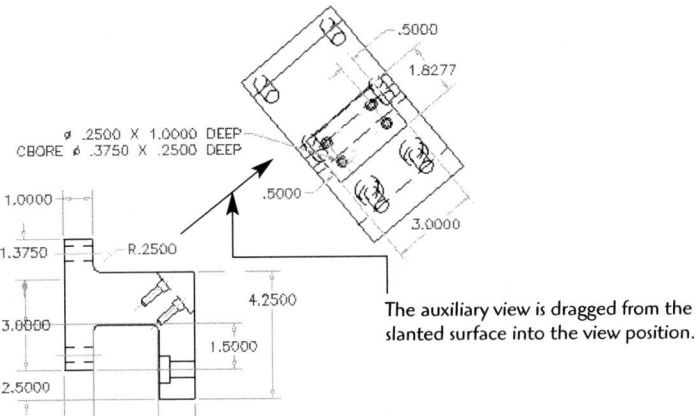

The auxiliary view is dragged from the slanted surface into the view position.

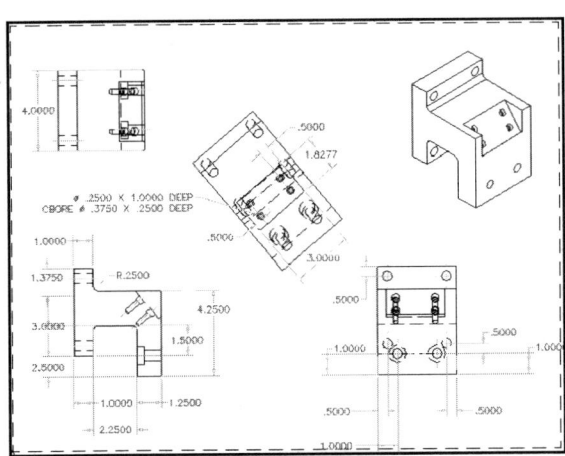

The completed auxiliary view incorporated into the existing orthographic drawing

Drawing Project 8–1

Detail Drawing Development from Models

Reopen your prevously produced models and convert these models into detail drawing orthographic views. Include a minimum of a top, a front, and side views. The drawing should be complete enough that a reader could produce the part. When you have finished, print out a copy of the drawing for your instructor.

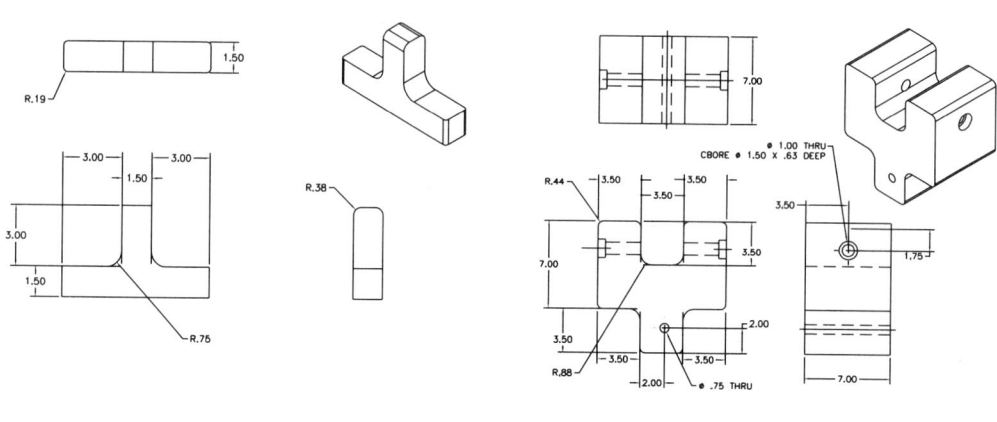

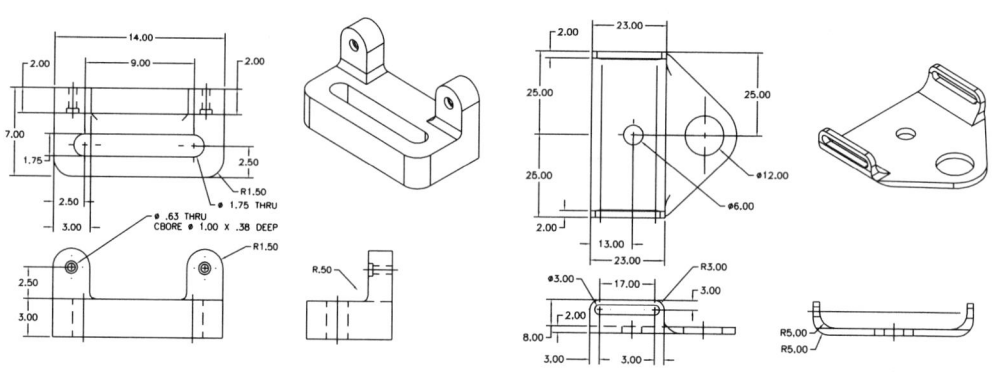

Drawing Project 8–1 *continued*

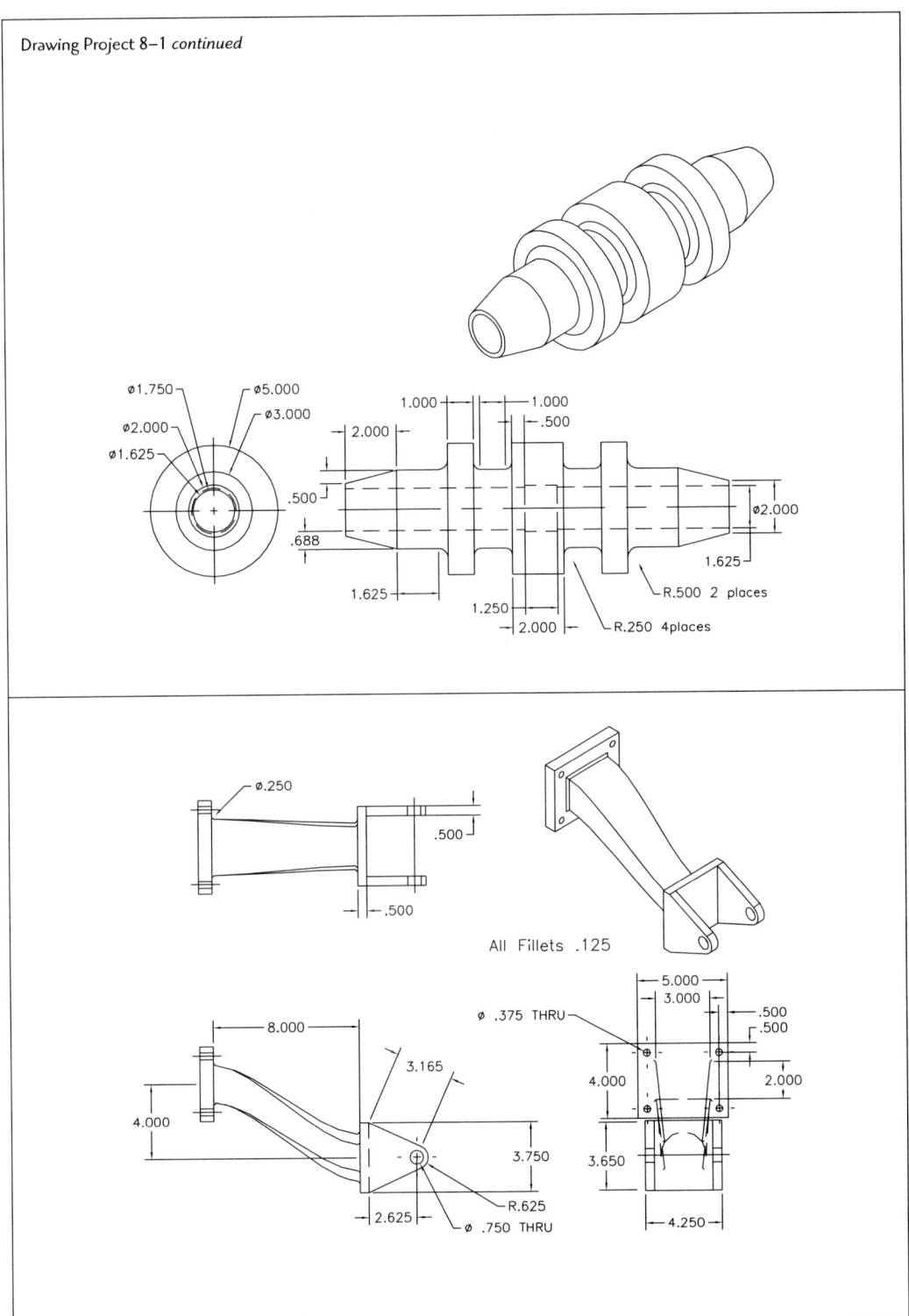

Drawing Project 8–2

Advanced Detail Drawing Development: Auxiliary Views

Convert the isometric drawing into an orthographic detail drawing. Include a minimum of a top, a front, a side, and an auxiliary view. The drawing should be complete enough that a reader could produce the part. When you have finished, print out a copy of the drawing for your instructor.

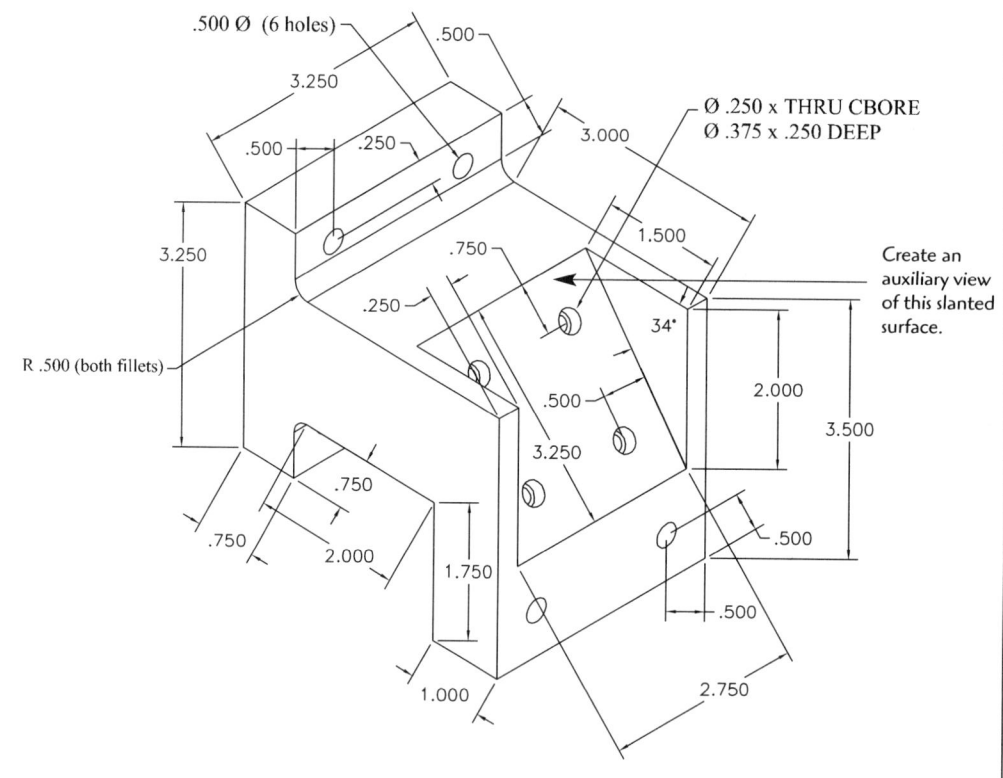

Drawing Project 8-3

Advanced Detail Drawing Development: Section View

Model the wheel disk shown. Include standard orthographic views and a section view of the wheel disk. The drawing should be complete enough that a reader could produce the part. When you have finished, print out a copy of the drawing for your instructor.

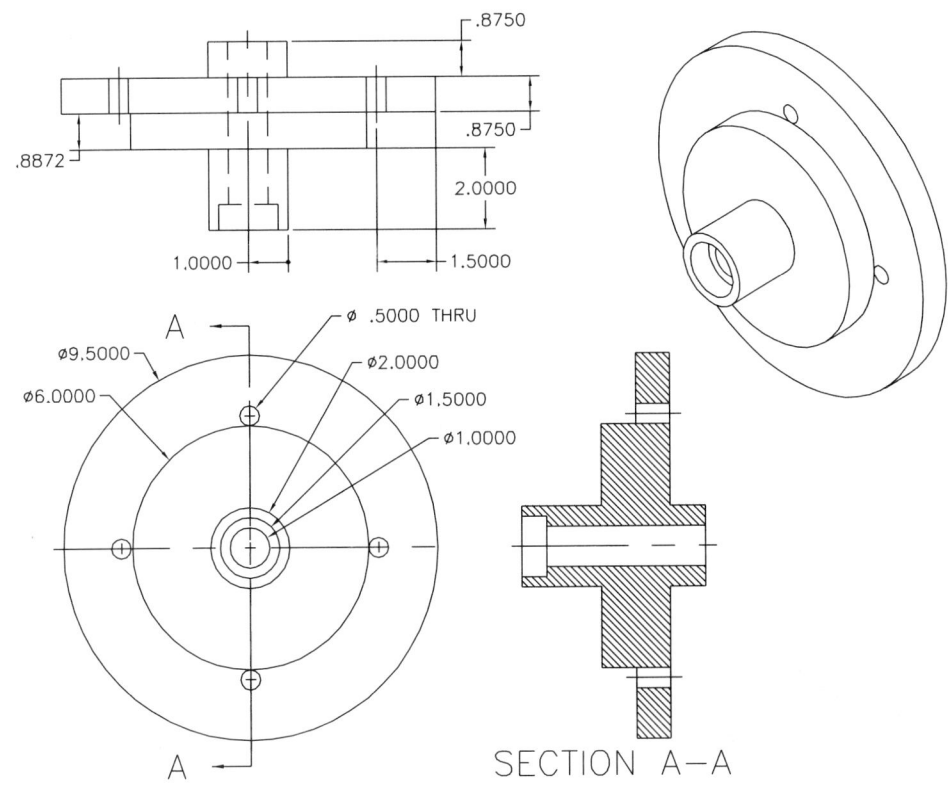

Drawing Project 8-4

Advanced Detail Drawing Development: Auxiliary View and Detail View

Model the angle plate shown. Include standard orthographic views and an auxiliary view of the slanted surface, and a detail view of the angle plate. The drawing should be complete enough that a reader could produce the part. When you have finished, print out a copy of the drawing for your instructor.

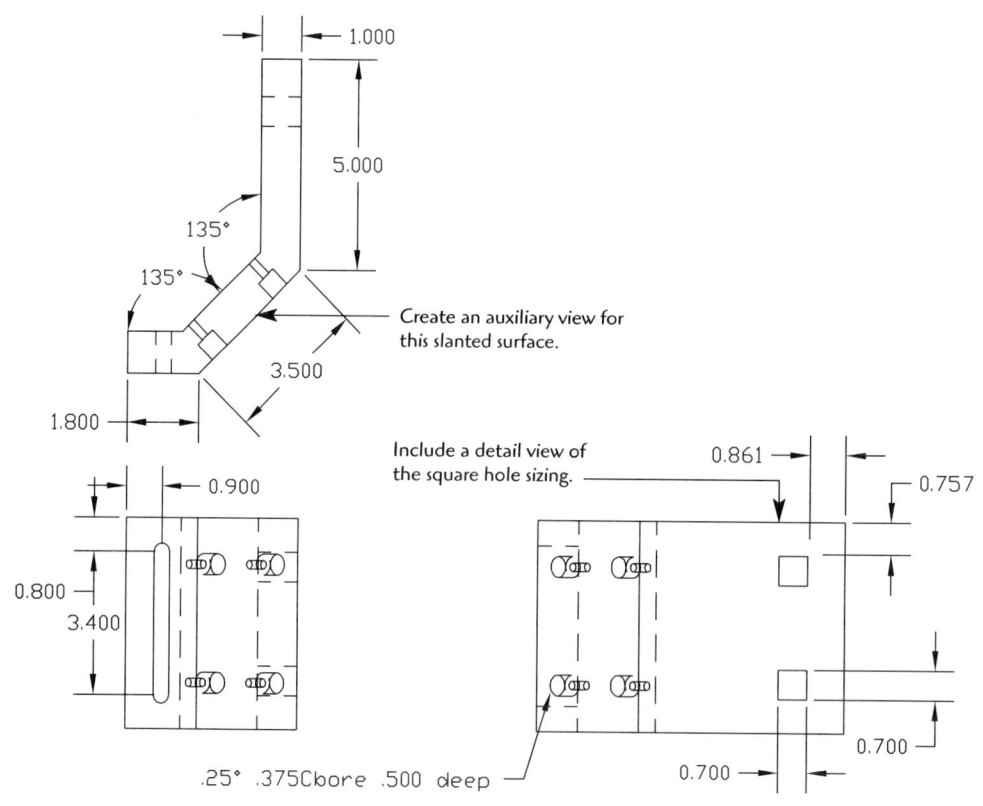

HELIX

The **Helix** command is actually a series of commands used together to create a helix. The 3D path is the key to the helix operation. The helix shape is useful for any number of mechanical models and assemblies that use springs, industrial shock absorbers, and the like.

Let's create a basic helix:

Step 1   Type **AMBASICPLANES** ↵
or click the icon.

Click on the screen to place the work planes. (This sets up the basic XYZ work planes.)

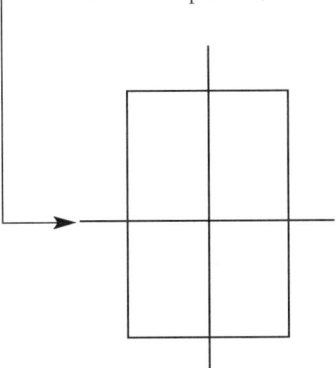

141

Step 2 Type 8 ↵.
(This displays the three basic planes in an Iso view.)
Make one of the planes the active sketch plane.

Type **AMSKPLN** ↵
or click the icon.

Click on the work plane and ↵↵.

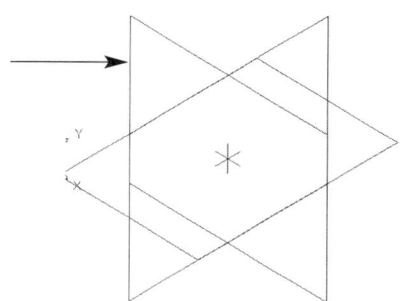

Step 3 Now, you need a work axis through the work planes to provide an axis for the center of the helix.

Type **AMWORKAXIS** ↵
or click the icon.

`Select cylinder, cone, torus or [Sketch]:`

Type **S** ↵.
(The S indicates you will sketch the work axis on the screen.)
Draw a two-point line on the current sketch plane. (You can use Osnaps or Ortho to keep the line straight and in the center of the work plane.)

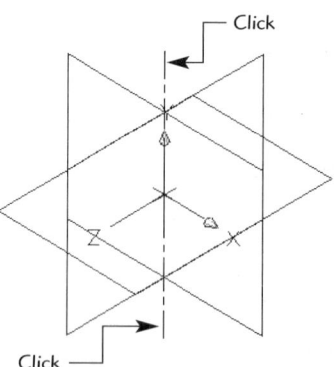

Step 4 You now need to set the work plane perpendicular to this work axis.

Type **AMSKPLN** ↵
or click the icon.

Click on the work plane and ↵↵.
— (Select the work plane at the base of the three work planes.)

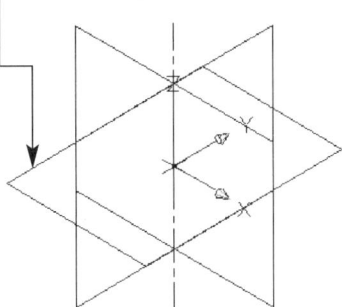

Step 5 Now, set up the helix path.
Type **AM3DPATH** ↵
or click the icon.

Depending on your version of Mechanical Desktop, you may be asked for the type of 3D path.
Type H ↵ (for helical).
Select the work axis. ——

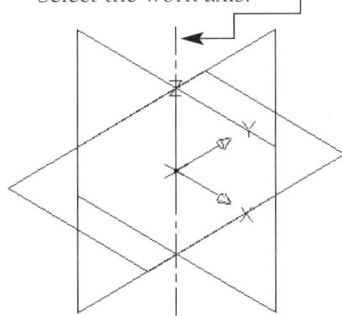

The Helix dialog box appears.
—— Set the **Revolution, Pitch,** and **Diameter.**
After making the settings, click **OK.**

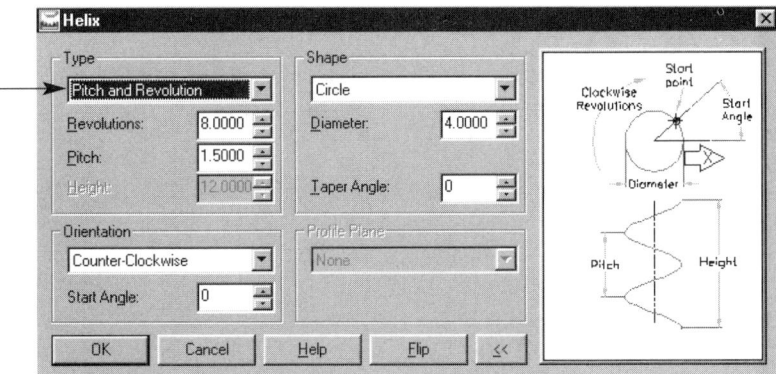

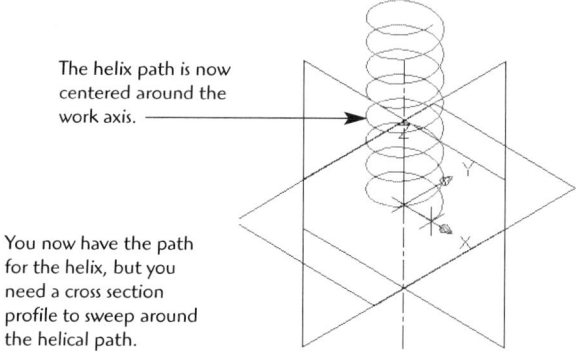

The helix path is now centered around the work axis.

You now have the path for the helix, but you need a cross section profile to sweep around the helical path.

Step 6 Due to the special angle of the helix, you need a special work plane to match up perpendicular to this endpoint.

Type **AMWORKPLN** ↵
or click the icon.

`Select Normal to Start:` *(This means to place a work plane at a right angle to, or normal to, the start point of the end of the helix.)*

Make sure the Create Sketch Plane box is checked.
Click **OK**.

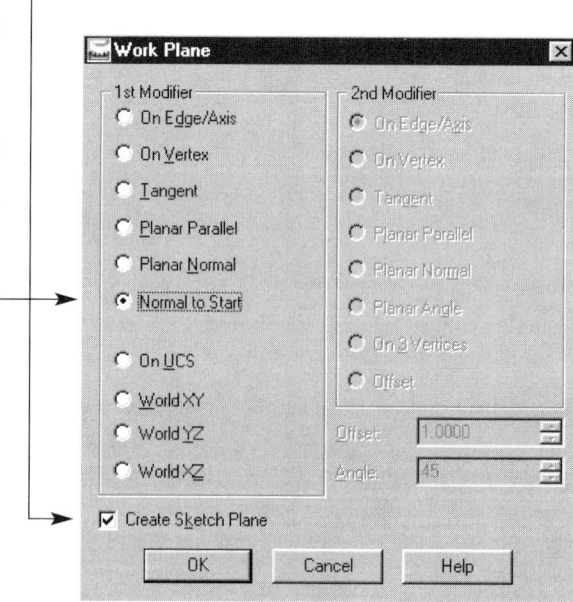

HELIX | 145

Select the path: *(Select the helix.)*

Press ↵.

(This confirms the new work plane/sketch plane placement.)

Although it looks awkward, this new work plane is perpendicular to the endpoint of the helix.

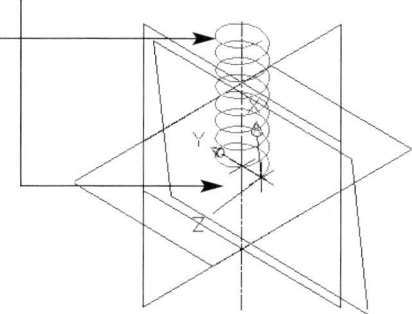

Step 7   You are now ready to sketch the cross section profile. You may draw this profile in the current Iso position, or in a flat plan view with the PLAN command. Be sure to Osnap the profile to the endpoint of the helix start point.

Let's use a circle for the profile.

Type **Circle** ↵.
Type **END** ↵.

Click directly on the end of the helix line (called the start point). Drag the circle to any size.

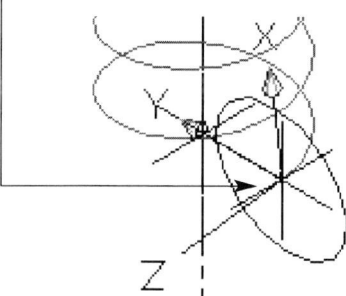

You must now fully constrain the cross section profile. First, profile the cross section profile.

Type **AMPROFILE** ↵
or click the icon

Select the circle ↵.

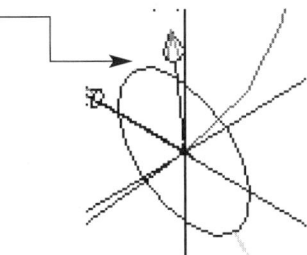

Now, you must fully constrain the cross section with constraining dimensions.

Type **AMPARDIM** ↵
or click the icon.

Select the circle, then away from the circle to place the dimension.

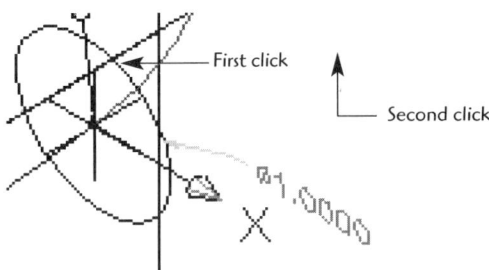

Finally! You are ready to sweep the profile around the 3D helix path.

Type **AMSWEEP** ↵
or click the icon.

— Use the **Path Only** option.
Click **OK.**

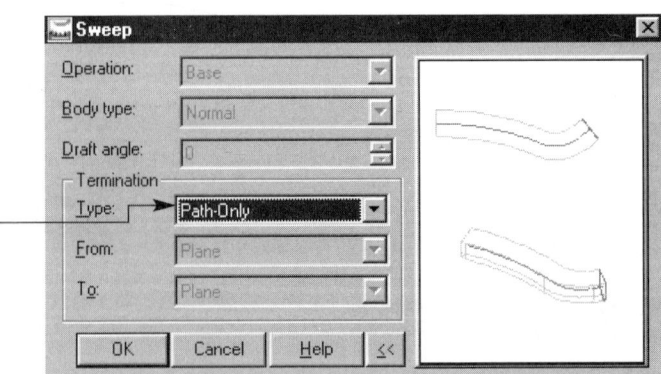

HELIX | 147

Be patient! The helix sweep process can take a while.
If the solid helix is not visible immediately, render or hide the view.

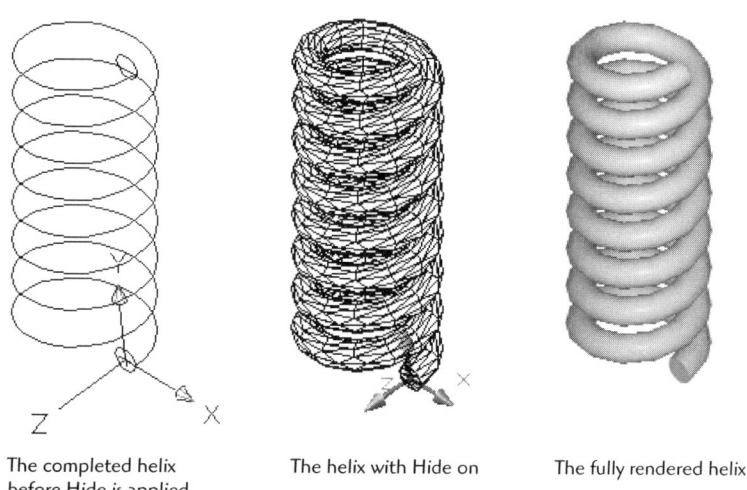

The completed helix before Hide is applied   The helix with Hide on   The fully rendered helix

For a Deeper Understanding about
Helix

There are a number of options that can be used to define the helix.

- Pitch and revolution
- Revolution and height
- Height and pitch
- Spiral

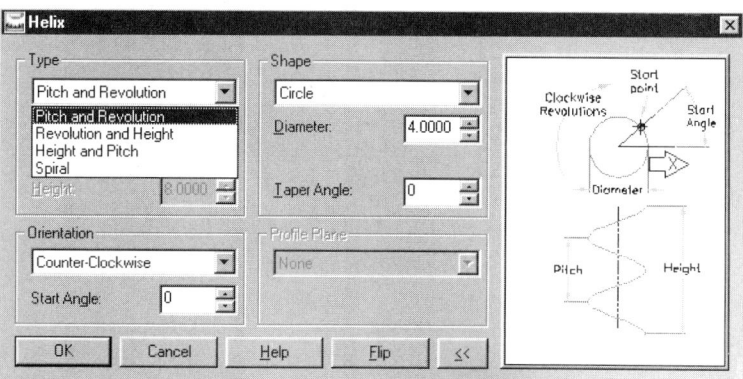

The **pitch** is the center-to-center spacing between the revolutions.
The **height** is the total height of the helix.

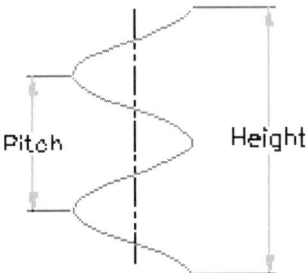

The **revolution** is the number of complete turns around the helix.
The **spiral** option starts from the center and moves out.

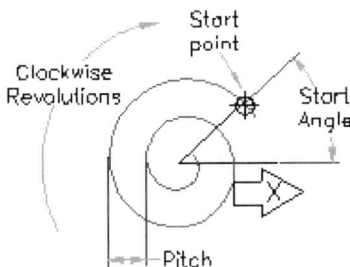

The Helix dialog box shows additional options.

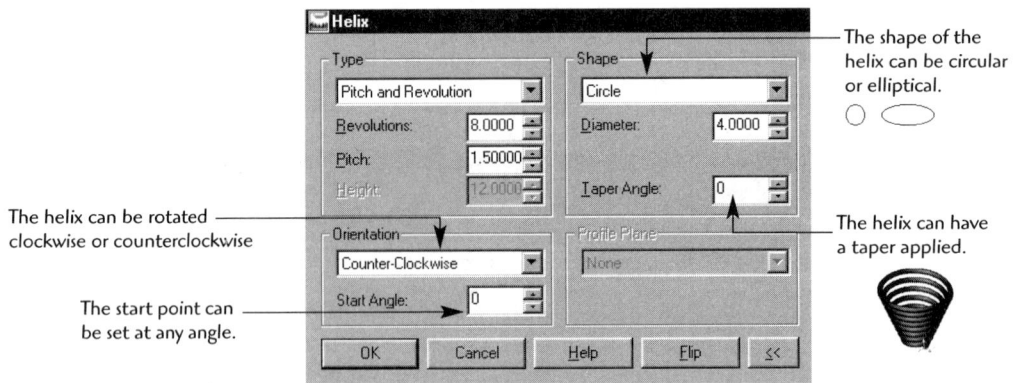

The shape of the helix can be circular or elliptical.

The helix can have a taper applied.

The helix can be rotated clockwise or counterclockwise

The start point can be set at any angle.

Drawing Project 9–1

Industrial Shock

Model the industrial shock to the exact dimensions indicated. Use the helix, and extrude to produce the part. Add the fillets, chamfers, and holes as needed. Create a 4 inch Helix diameter, and a .5 inch profile. When you have finished, save the drawing and click **8** for an isometric view. Print out a copy for your instructor.

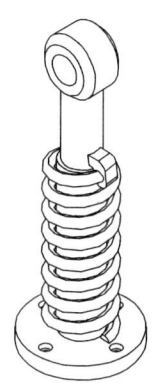

DETAIL A
SCALE 1:3.33333

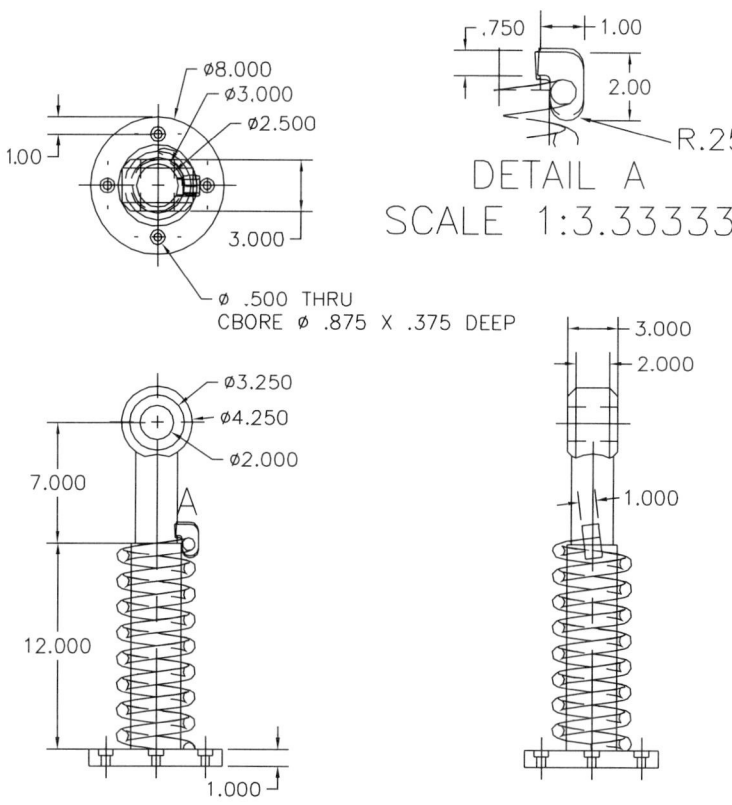

PARAMETRICS

CREATING PARAMETERS IN A MODEL

Now, let's create a basic part with parameters so that the basic shape can be reused but in a variety of usable sizes. For example, what if you need to manufacture a part in a basic shape, but the same basic shape is needed in 8″, 12″, 16″, and possibly more sizes. You need to create only one model, but you will add parameters to the model so you can create the other required sizes without redrawing the model each time.

Although parameters are easy to use, planning is required. You must engineer the correct sizes and ratios into the model. For example, if a mounting bolt hole is required to be two times the thickness of the mounting plate, you can program in a formula to cause these ratios to stay consistent whatever the sizes are. Instead of using a diameter size on the holes, you would use the formula **Diameter = thickness $\times$ 2** (d1 = T1 $\times$ 2).

Some design codes require a minimum radius on parts to prevent corner tearing or cracking under stress. If the requirement is that the minimum radius be no less than 10% of the part length, you might use a radius dimension of **Radius = Length $\times$ .10** (R1 = L1 $\times$ .10). Each time a radius dimension is required, its size will be 10% of the length dimension.

151

CHAPTER 10

Let's try an example.

Step 1 Create the basic shape of this tension rod using plines. Be sure to close the shape. (You will add fillets and holes later.)

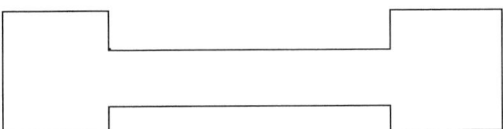

Step 2 Perform the initial constraining step with the AMPROFILE command.

Step 3 Before fully constraining with dimensions, set the dimensions to display as equations. From the pull-down menus select:

Part

> Dimension > Dimension as Equation

or click the icon.

Note, first, how the automatic constraining is applied to the shape: (When you profile the shape these constraints are applied.)

V indicates *vertical* constraints (line must stay vertical).

H indicates *horizontal* constraints (line must stay horizontal).

C indicates *collinear* constraints (lines or curves must stay in a collinear plane).

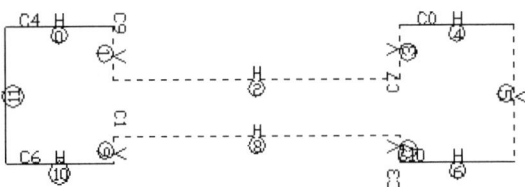

Step 4 Now, you will fully constrain with equations and dimensions.

Use AMPARDIM or click the icon and click the line to be dimensioned, then click away from the line to place the dimension. Use **2** as the first constraining dimension.

The first constraining dimension, d0, is set to 2.

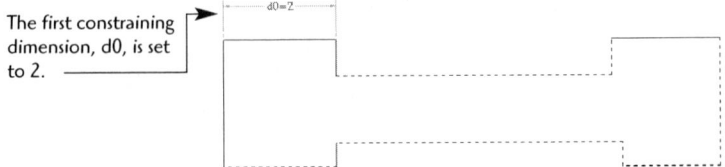

Step 5 Using the same dimensioning procedure, continue to add dimensions and formulas to the profile. Since you want the perpendicular interior line to always remain one-third the length of the top line, dimensionally constrain it as "d0 divided by 3" (enter it as **=d0/3 ↵**).

Note that each dimension added has a *dimensional name* such as d0 and d1. These are variable names that allow you to reference the parametric dimension without assigning a permanent dimensional value. In other words, it allows you to use the equations.

Step 6 Now, create a constraint for the left vertical side and make it three times the d1 value (entered as **=d1*3 ↵**).

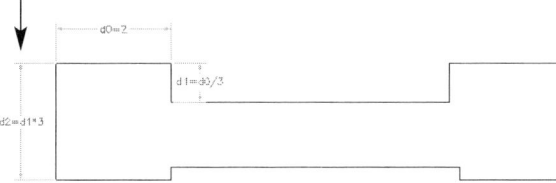

Note how you can make one dimensional value or equation *reliant on another* for its value. The d2 value is always three times the d1 value, whatever its value becomes.

Step 7 Continuing the constraining, make the vertical side opposite dimension d1, equal to d1 (entered as **=d1 ↵**).

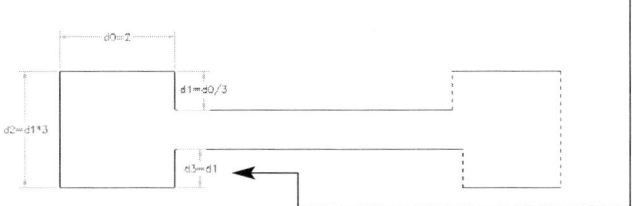

In attempting to constrain an edge you may get the message "**Adding this dimension would over constrain the sketch.**" When constraining, you must always have a flexible side to allow for changing dimensional values. If you attempt a constraint that will lock in a profile without this flexibility, you will get this warning.

So, what should you do if you get the warning? If you do get the warning, you may not actually need the constraint at that side, or you may simply try to add a constraint to another side that will allow the same results. The other option is to try different constraining combinations. Usually, more that one style of constraining layout will work for a design.

Step 8   Add two constraints to the opposite side. These constraints will keep the left side the same as the right side.

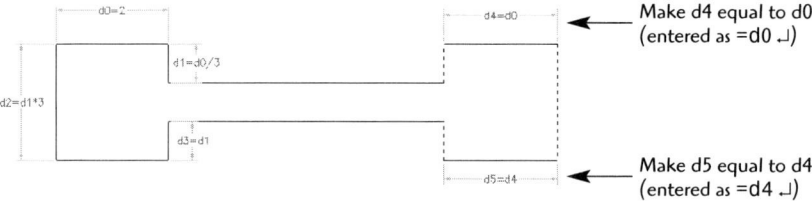

Make d4 equal to d0 (entered as =d0 ↵)

Make d5 equal to d4 (entered as =d4 ↵)

Step 9   Add one more constraint to the profile to control the narrow section length. Keep the d6 length three times as long as the d0 length (entered as =d0*3 ↵).

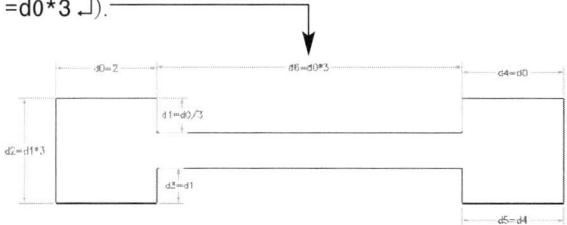

On the command line you will see the message "Solved fully constrained sketch." This indicates that all possible dimensional constraints have been applied to the profile sketch. It is not always necessary to fully constrain the sketch, but fully constraining allows for maximum parametric flexibility. A fully constrained profile can be changed and adjusted to all possibilities. A partially constrained sketch will have limitations on what dimensional parameters can be adjusted.

Step 10   Turning the profile sketch into a 3D feature automatically constrains the thickness added to the model. Equations can be used just as dimensions can be applied to the feature creation process. You will extrude the profile using an equation that will control the thickness to a percentage of the width of the part. The thickness will be 60% of the width of the part (entered as =d2*.60 ↵).

PARAMETRICS | 155

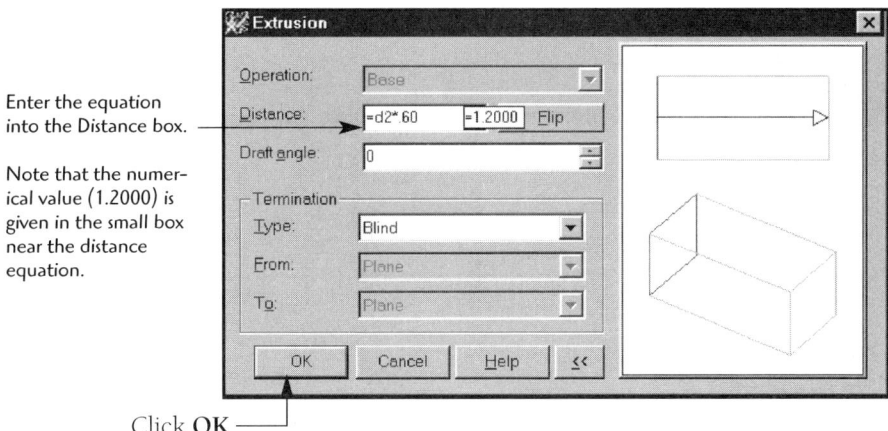

Enter the equation into the Distance box.

Note that the numerical value (1.2000) is given in the small box near the distance equation.

Click **OK**.

Step 11 Click 8 ↵ (to give you an iso view of the part).

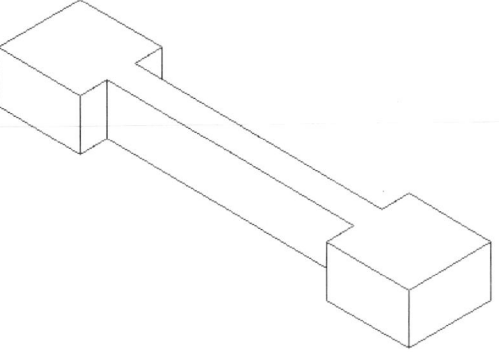

Step 12 Continue modeling by adding the holes. You will use an equation for the holes to keep the sizes appropriate for the part size. Make the hole diameter 40% (.40) of the part width (entered as =d0*.40 ↵).

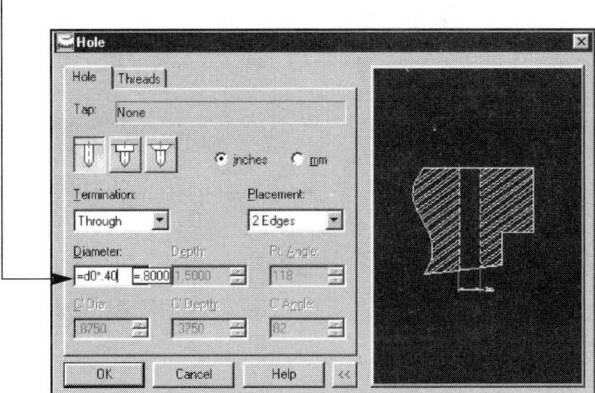

You want the location of the holes to stay centered, so you will use an equation that divides the width by two whatever its size happens to be (entered as **=d0/2** ↵).

Drag into approximate hole position.

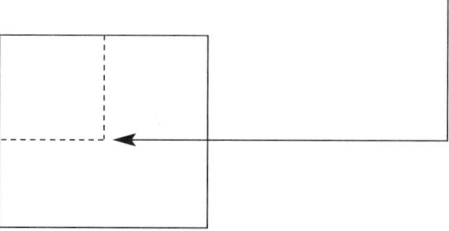

```
Enter distance from first edge =d0/2 ↵
Enter distance from second edge =d0/2 ↵
```

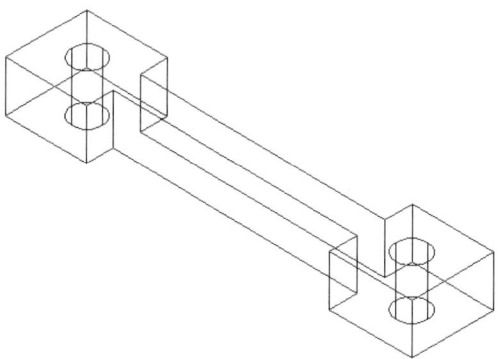

Step 13 Add the fillets. Again, you need to keep the fillets proportional to the size of the part. Make the radius equal to one-third the width of the part (entered as **=d1** since d1=d0/3).

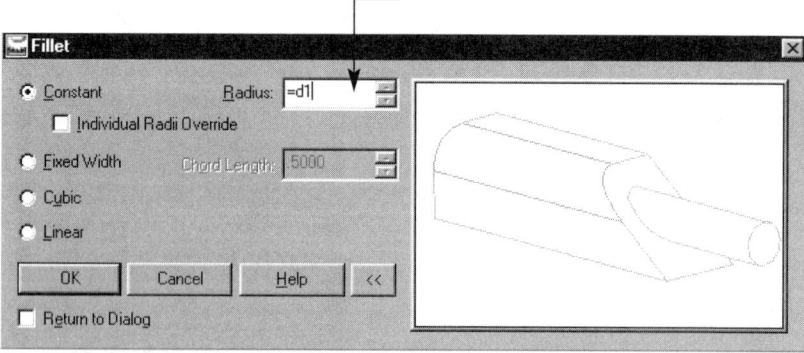

Add fillets to all the corners of the part.

Now, let's do something powerful with the parametric model we have created. Make more models by changing only one dimensional value!

Step 14 You will proportionally change the model to a different size by changing only the value of d0. The preferred method for editing your model is in the desktop browser.

Right-click on **Extrusion Blind** and select **Edit Sketch**.

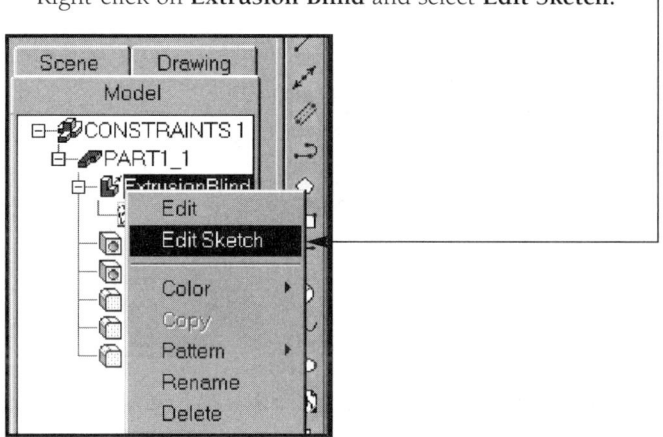

The sketch reappears.

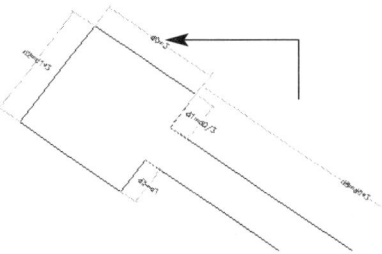

Type **AMMODDIM** ↵
or click the icon.

Select the d0 value.

On the command line, change the value of d0 to 3 ↵.
Now, you need to update the changes you have made.

Step 15 Type **AMUPDATE** ↵↵.

Step 16 Try changing the value of d0 so you can create different proportional sizes of your model without having to redraw it each time. In the figure, d0 is set to 2, 4, and 8.

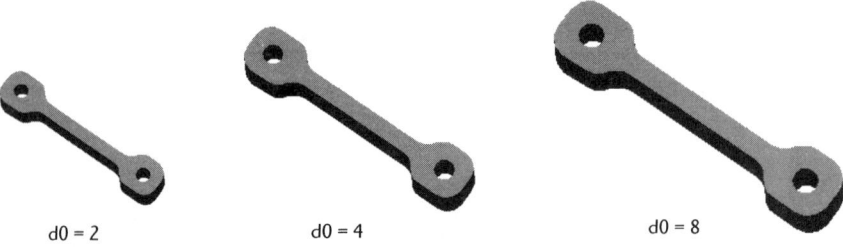

d0 = 2   d0 = 4   d0 = 8

For a Deeper Understanding about
Parametric Equations

The following are helpful hints for adding equations to a parametric model:

- Always start an equation with "=" (the "equals" symbol).
- Try to constrain the most important dimensional sides or area first.
- Don't be afraid to reconstrain the dimensional constraints if the first layout does not work. (Save the profiled sketch under a different name before applying the dimensional constraints. You can then reopen the original to reapply new dimensional constraints)

- Use "fix point" if the profile sketch moves in the wrong direction when dimensional constraints are added. Click the lock icon to activate fix point. Place the marker on the point to be locked.

Fix point marker

- Some constraining parameters can be set with the **Mechanical Options**. Click the Mechanical Options icon to bring them up.

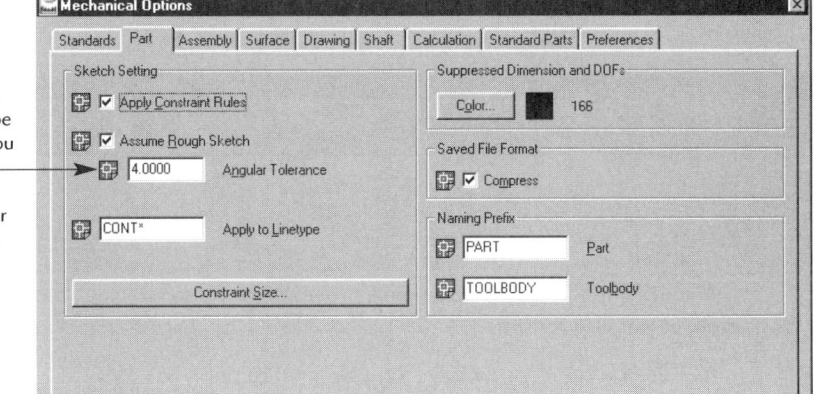

When the rough sketch is profiled, small angles will be straightened. If you need these small angles to remain, adjust the Angular Tolerance setting.

Math Options
- Use * (the **asterisk**) for multiplication.
- Use **/** (the forward slash) for division.
- Use **+** and **-** for addition and subtraction respectively
- Use **^2, ^3, ^4,** etc., for exponents.
- Use **sqrt** for square root.
- Use **pi** for pi (3.14).
- Use **sin, cos, tan, asin, acos, atan,** etc., for trigonometric functions.
- Use **log** for logarithmic functions.
- Other math functions are available for use. Check your user reference manual.

Constraints can be displayed three different ways: (From the pull-down menu click Part, Dimensioning.)

Di̲mensions As Parameters

Dimensions As N̲umbers

Dimensions As E̲quations

| Display as **parameters** | Display as **numbers** | Display as **equations** |

Display As Equations is a very helpful setting, since both the value and the parameter name are displayed.

Recall that the following 2D sketch constraints can be used to constrain the sketch.

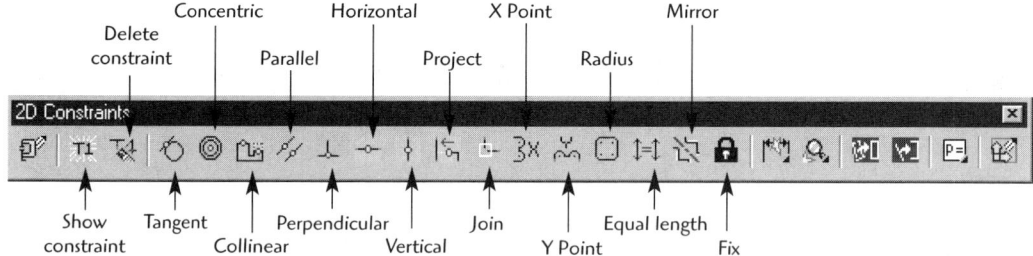

Most of the 2D sketch constraints are applied the same way. The following example applies a parallel constraint:

- Click on the constraining icon you need to apply.

- Select the first line or object.

- Select second line or object.

The parallel line constraint is displayed on the two lines.

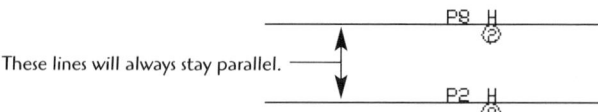

These lines will always stay parallel.

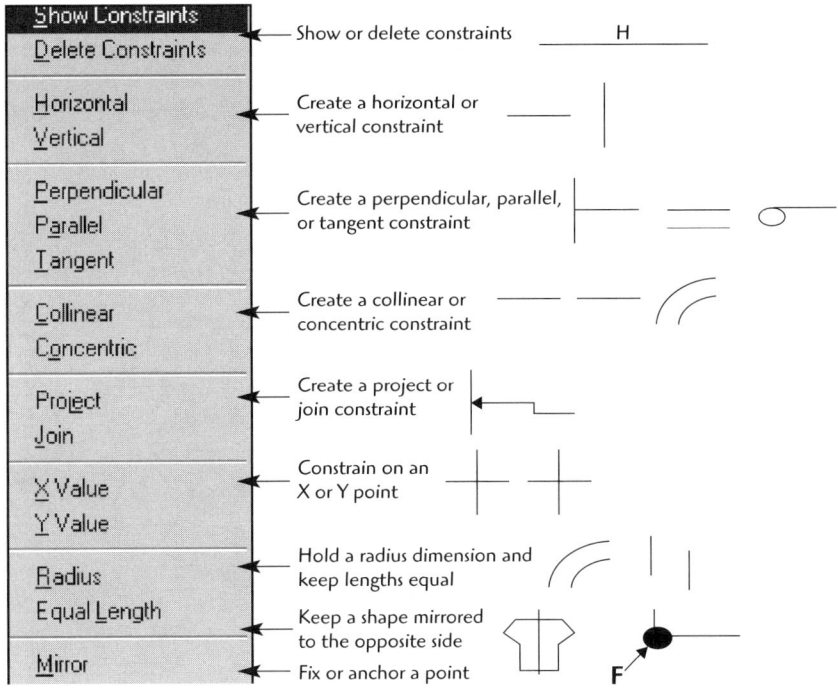

PARAMETRIC MODELING WITH TABLE-DRIVEN VARIABLES

Mechanical Desktop allows you to use **external table data** to control size variables and equations for a model. Think of how a part might be ordered for a project. One part is created, but the table or spreadsheet lists all available sizes for the part. Whatever size is requested is applied to the model for the order.

The next practice project will use an Excel® spreadsheet to control size and equation data for the model. You will first enter the variable data, then enter the data into the model. The model will be a basic shear pin design of five variable sizes. You will create only one shear pin, and the table will generate the other five sizes.

Step 1 To set up the variables, type **AMVARS** ↵ or click the icon.

Step 2 Click **New**.

Input the **Name, Equation,** and **Comment,** in the New Part Variable box.

Click **OK**.

Repeat for as many variables as necessary. (Just click New and add the other variables.)

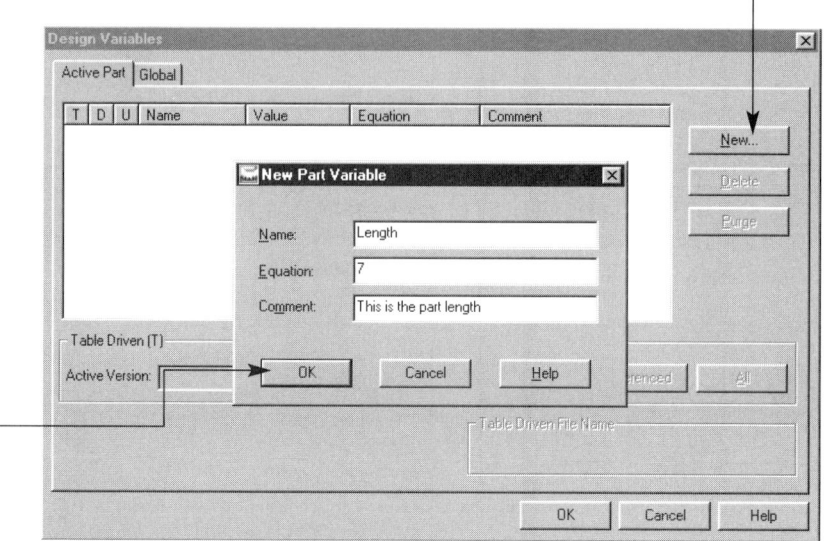

Create a design variable table with the following variables: (Repeat step 2 for these additional variables.)

- **Length** (length of the pin)
- **Hdiameter** (head diameter of the pin)
- **Sdiameter** (shaft diameter of the pin)
- **Plock** (pin lock hole)
- **Hwidth** (head width)

Your table should look similar to the figure. For the equation, use the initial sizes listed.

Step 3   Draw and profile the part.

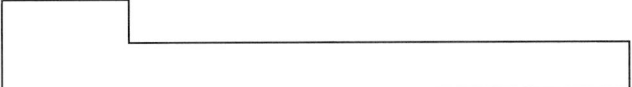

Step 4   Now, fully constrain the part with AMPARDIM, but when prompted for the dimension value, **add the name of the variable** instead of the dimension. Continue this process with each dimensional constraint. (Note in the figure that the dimension you added for the equation becomes the distance for each variable.)

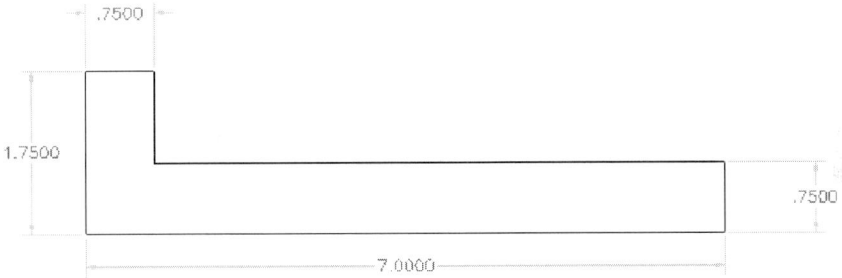

If you need to edit any of the variables, type AMVARS ↵ again and click on the **Global** tab. (All variables should appear under global.) Double-click on any of the variable properties and key in the changes.

Step 5   Now, revolve the profile, and add the block hole (pin lock hole).

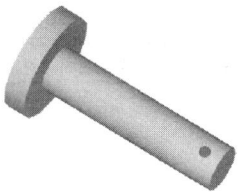

Step 6 You are now ready to send the dimensional data to the external spreadsheet. If you're not already in the Design Variables dialog box, type **AMVARS** ↵. (Make sure the Global tab is selected.)

Click **Setup** then **Create**. (Create is on the top right side of the next box that will appear.)

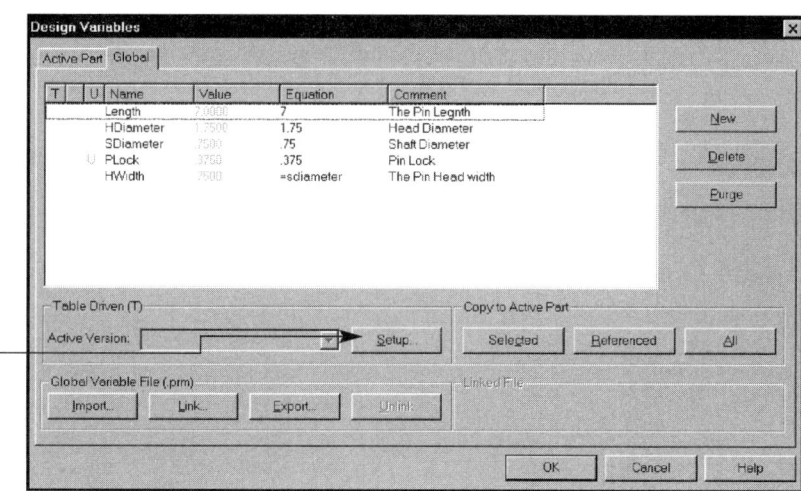

Step 7 Enter the following path and file name: **C:\Program Files\Mechanical\NAME.XLS**. You can use any file path as long as you remember where the file is going. The name should be the same as your drawing name, and .XLS Microsoft Excel extension. Click **Save**.

The Excel program will start.

This is the default layout for the spreadsheet. The first part you have created is named "Generic."

	A	B	C	D	E	F
1		Length	HDiameter	SDiameter	PLock	HWidth
2	Generic	7	1.75	.75	.375	=sdiameter
3						
4						
5						
6						
7						

Step 8 The table names and values can be adjusted and added to. Adjust the table for your requirements by changing the part names and adding some new parts and sizes.

Click into the cells and add the new pin sizes. Any practical sizes can be added.

	A	B	C	D	E	F
1		Length	HDiameter	SDiameter	PLock	HWidth
2	Pin 1	7	1.75	.75	.375	=sdiameter
3	Pin 2	9	2	1	0.5	=sdiameter
4	Pin 3	11	2.25	1.25	0.625	=sdiameter
5	Pin 4	14	2.25	1.5	0.75	=sdiameter
6	Pin 5	16	2.5	1.75	0.875	=sdiameter
7						

Here four pins have been added, with a new size for each new pin.

Step 9 To save the Excel table click:

File

➢ Save

Exit out of the Excel spreadsheet table by clicking the X in the upper right corner.

The Table Driven Setup dialog box should still be displayed. (If it is not, click AMVARS, then Setup.)

Click **Update Link.**

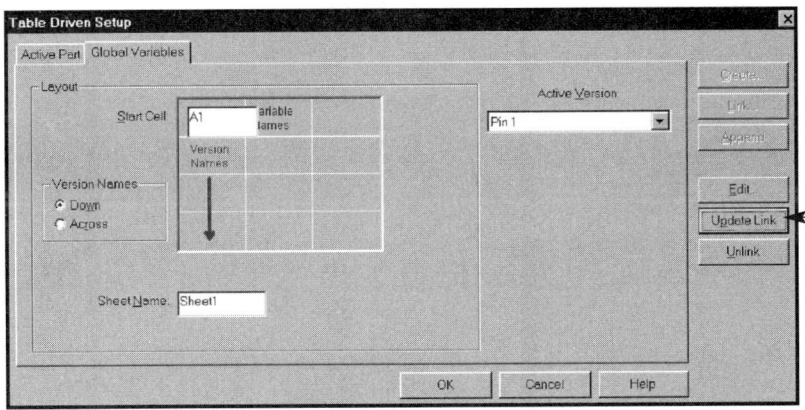

Click **OK,** and **OK** again.
You are now back out to the model page.

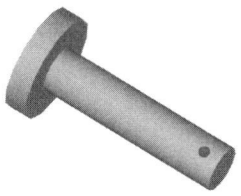

Now activate the sizes you have created by double-clicking on each pin setting in the Global Table list of the desktop browser.

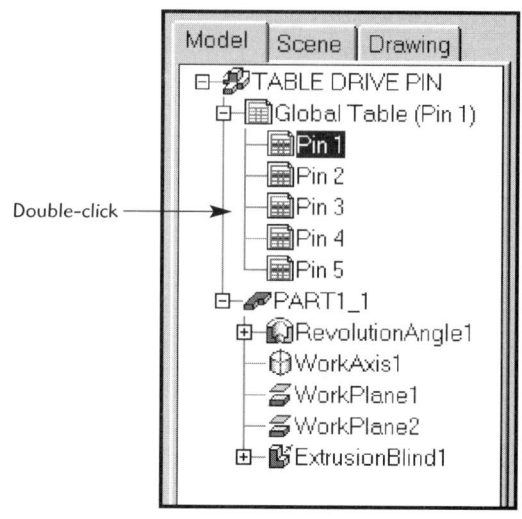

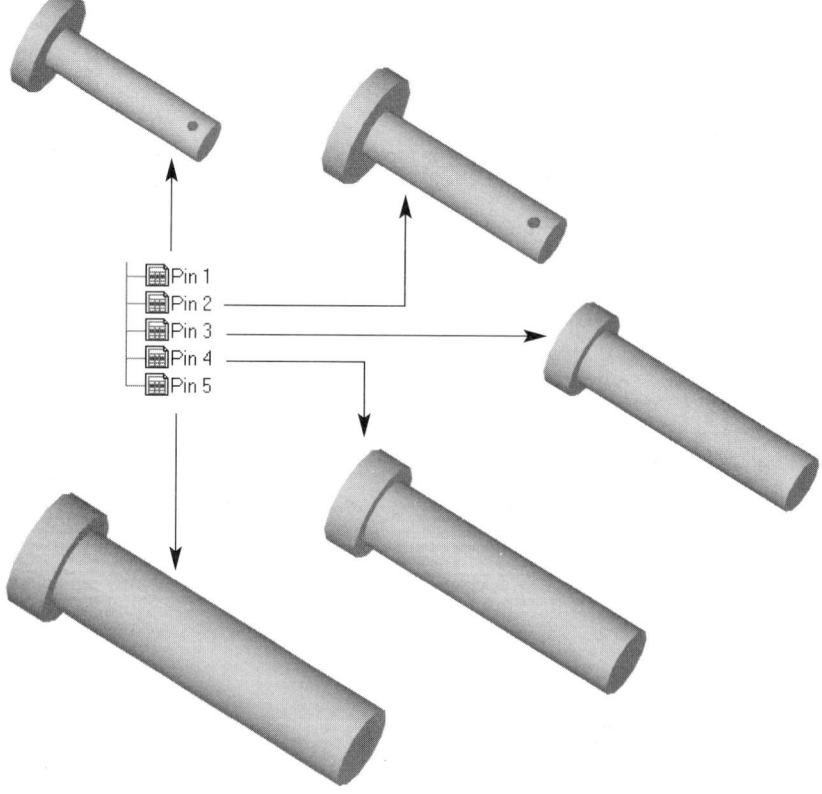

What about the orthographic drawing views of these models, will they change and update? Yes! Create orthographic views for one of the models, then double-click on one of the pin sizes in the Global Table of the desktop browser. You will see the new sizes applied to the orthographic views in the drawing display.

The Excel table data (below) is inserted into the corresponding variable size on the model for each pin size.

	Length	HDiameter	SDiameter	PLock	HWidth
Pin 1	7	1.75	.75	.375	=sdiameter
Pin 2	9	2	1	0.5	=sdiameter
Pin 3	11	2.25	1.25	0.625	=sdiameter
Pin 4	14	2.25	1.5	0.75	=sdiameter
Pin 5	16	2.5	1.75	0.875	=sdiameter

Note the size adjustment to the orthographic views as the pin model sizes are changed.

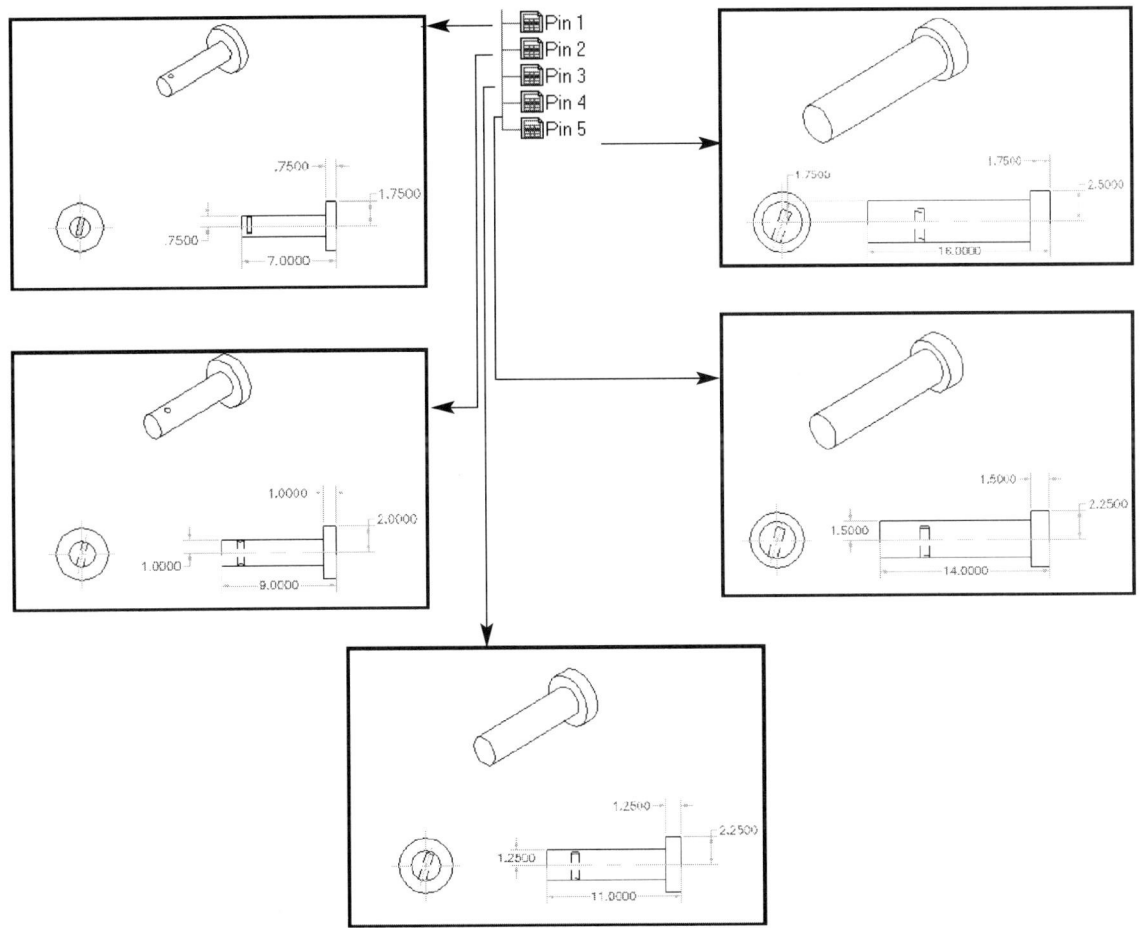

Drawing Project 10-1

Basic Corner Bracket

Model one basic corner bracket including complete parameters so the size can be easily adjusted, then create three more corner bracket sizes. Make the second bracket half the original size, the third twice the original size, and the fourth 30% (.300) the original size. When you have finished, produce and print out orthographic detail drawings of each of the four basic corner bracket sizes. Turn in all four drawings to your instructor.

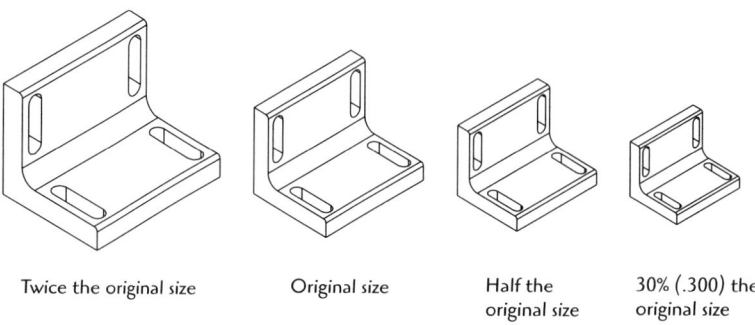

Twice the original size   Original size   Half the original size   30% (.300) the original size

Original Basic Corner Bracket

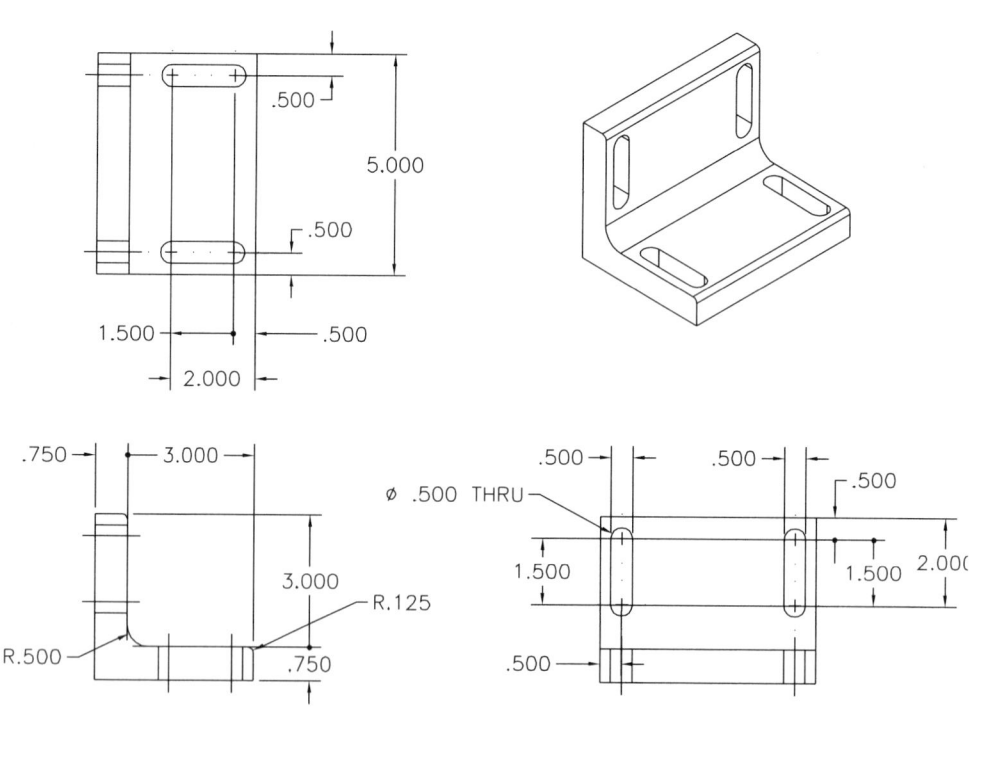

Drawing Project 10-2

Adjustable Bearing Bracket

Model one adjustable bearing bracket including complete parameters so the size can be easily adjusted, then create two more adjustable bearing bracket sizes. Make the second bracket 25% larger than the original size, the third 50% larger than the original size. When you have finished, produce and print out orthographic detail drawings of each of the three adjustable bearing bracket sizes. Turn in all three drawings to your instructor.

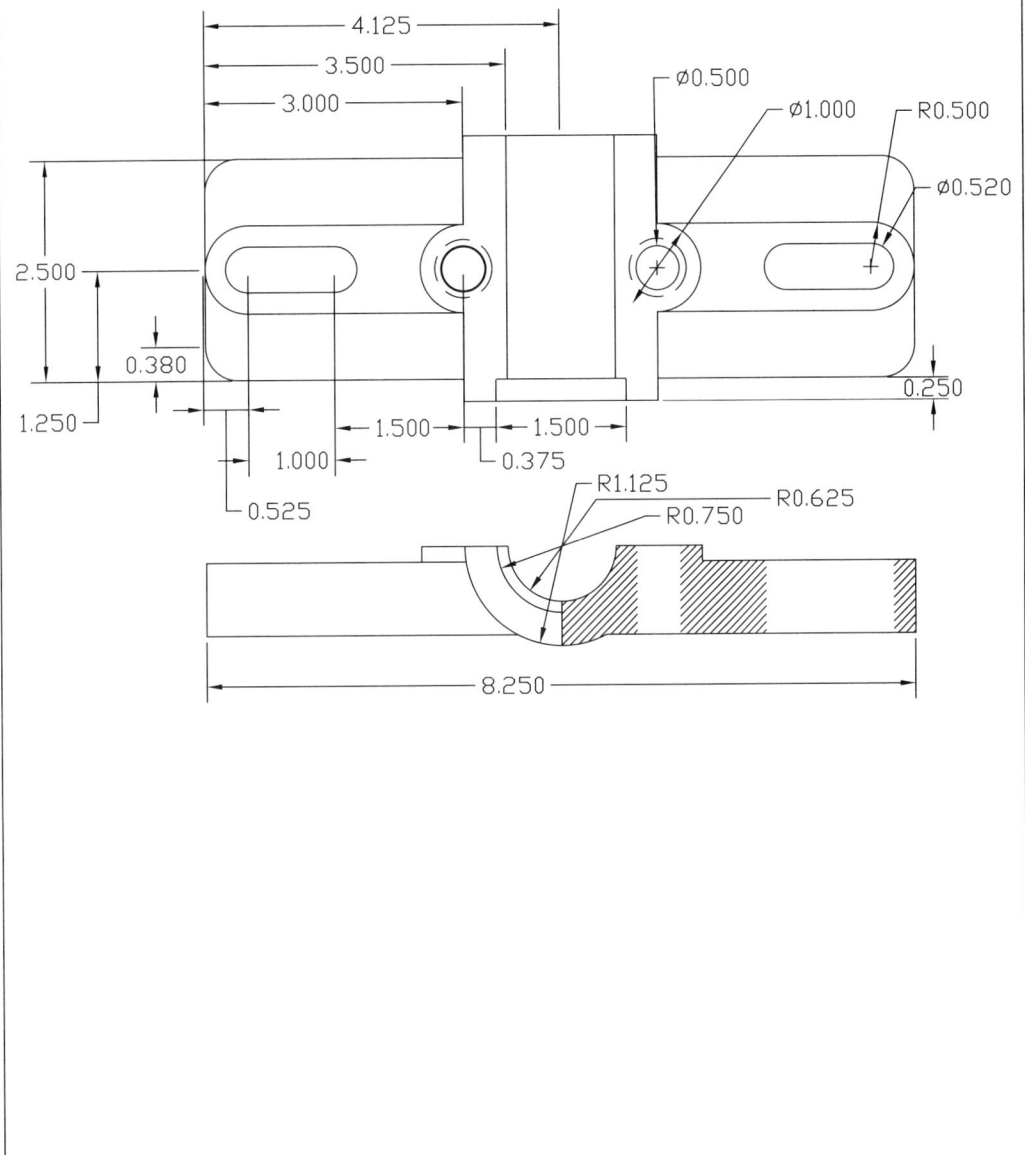

Drawing Project 10-3

Database-Driven Parts Project

Your company makes pulleys for the automotive and aerospace industries. You make five sizes depending on customer needs.

Create the following part with the four variable sizes. Draw the part only *one time* with the necessary variables. Enable the database to adjust to the necessary customer size needs. Create and print out a detailed orthographic drawing of each size. Turn your drawings in to your instructor.

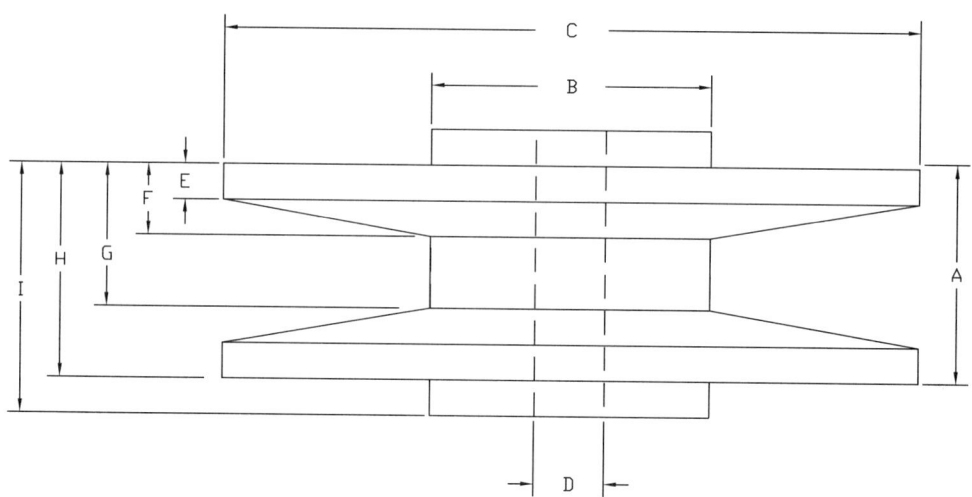

A	3	4.5	9	13.5	
B	4	6	12	18	
C	10	15	30	45	
D	1	1.5	3	4.5	
E	.5	.75	1.5	2.25	
F	1	1.5	3	4.5	
G	2	3	6	9	
H	3	4.5	9	13.5	
I	4	6	12	18	

Drawing Project 10-4

Support Arm

Model one support arm including complete parameters so the size can be easily adjusted, then create a database table with at least four additional sizes. Make two sizes larger, and two sizes smaller. You can engineer your own sizes for this project. (Remember, all sizes are in millimeters for this project.) When you have finished, produce and print out orthographic detail drawings of at least three of the table-produced sizes. Also print a copy of the database table. Turn in all drawings and the table to your instructor.

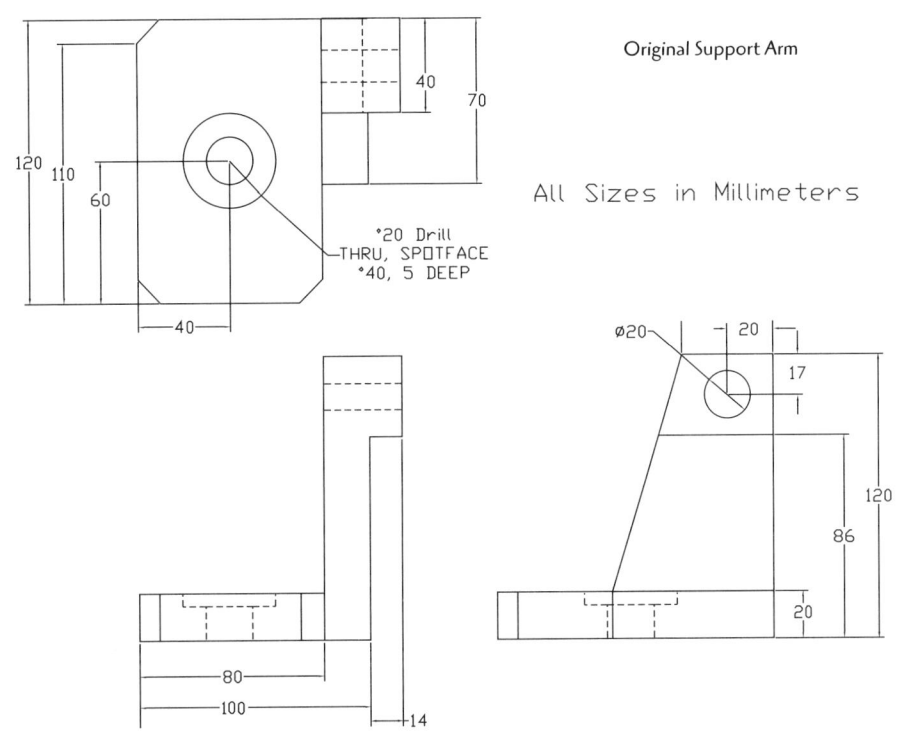

Original Support Arm

All Sizes in Millimeters

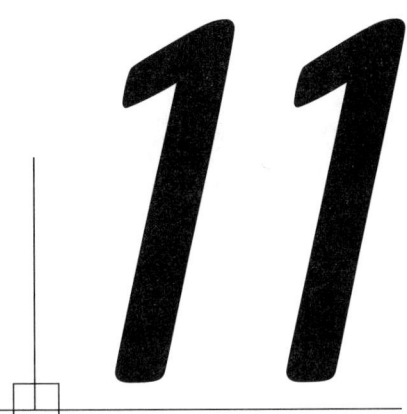

ASSEMBLIES

Mechanical Desktop offers much more than a parametric part designing tool; it also supports assemblies and analysis of these assemblies. Additionally, you can create assembly views of your final assemblies.

Let's look at an overview of the basic steps for creating assemblies:

1. Create each part that makes up the assembly. You must **identify** to Mechanical Desktop each time that a **new part** is being started.
2. Put each part into the **catalog.**
3. **Insert** catalog parts into the modeling screen.
4. Add assembly constraints between parts to define how they join together.
5. Test interferences between the parts.
6. Create exploded and assembly views in the scene mode.
7. Create an orthographic drawing of the entire assembly, as well as detail views where needed.

The following is a basic tutorial on creating an assembly.

Step 1 Create an initial part.

Type **AMNEW** ↵ and type **P** (for part) ↵
or click the icon.

Type in the name of the first part ↵.
(**Part 1** is the default name.)

Create and fully constrain the first part in the assembly.

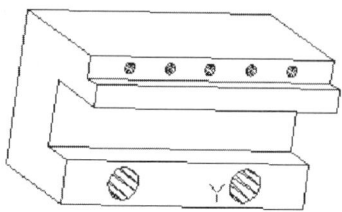

Step 2   Type **AMNEW** ↵ and type **P** (for part) ↵ or click the icon.

Type in the name of the second part ↵. (**Part 2** is the default name.)

Create and fully constrain the second part in the assembly.

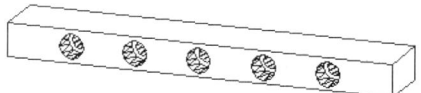

Continue to create each part needed for the assembly as in steps 1 and 2.

If new parts are getting in the way of one another, erase the parts that have been completed. When you erase, you will be prompted to remove it from the part definition. Click **No**. The part remains in the catalog, so you can recall it when you are ready to build the assembly.

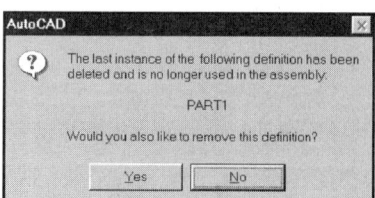

Each part of the assembly is now completed. The parts are also in the catalog. The parts are automatically put into the catalog every time you create a new part with the AMNEW command. Your next step will be to bring the parts out of the catalog and onto the Model screen for assembly and assembly constraining. Erase any parts still on the screen and bring in "fresh" parts from the catalog. (Again, remember to click **No** when prompted to remove the part from the definition.)

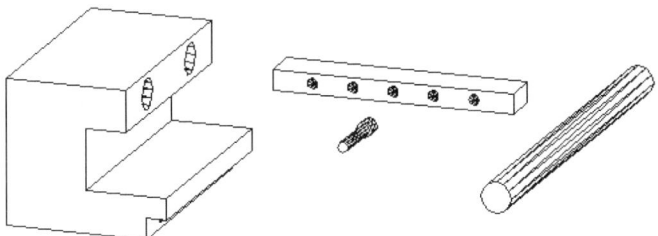

All the parts for the assembly have been created.

Step 3   Bring in parts from the catalog for assembly constraining.

Type **AMCATALOG** ↵
or click the icon.

The Assembly Catalog dialog box appears.

Click on the **All** tab to see the parts you have created. The External tab is used if you have outside parts to bring into the assembly.

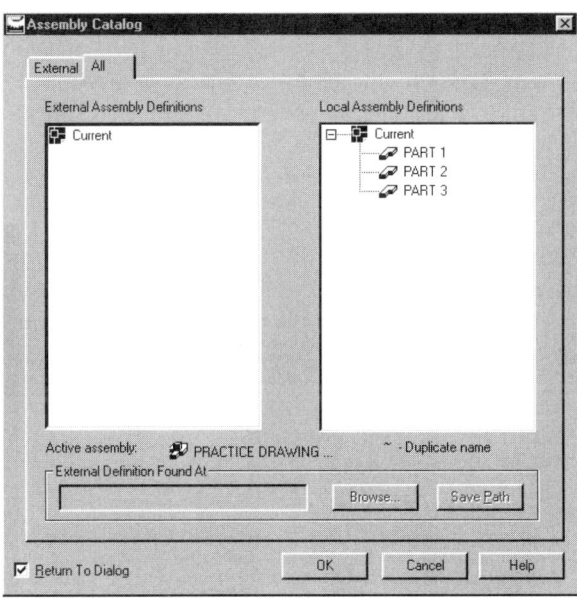

Double-click on a part in the catalog to bring it into the Model screen. Click the select (left mouse) button to place as many instances of the part as required for the final assembly.

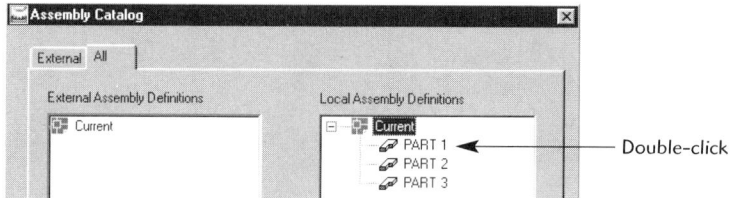

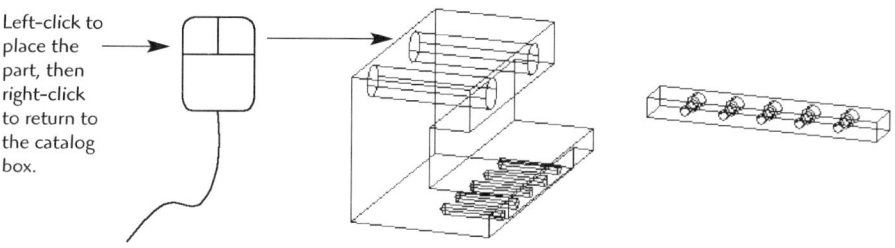

Continue to double-click and insert the parts that you need in the assembly. When you have finished placing parts in the Model screen, click **OK** in the catalog box. Do *not* click Esc or Cancel because all placed parts will also be removed from the Model screen. (*Hint:* If you prefer to have your screen less cluttered, bring in only a few parts at a time as you need them for assembly constraining.)

Step 4 **Assembly constraining** is the process of defining how each part will fit with respect to the other parts. Assembly constraining limits the degrees of freedom (DOF) of movement for each part. It also defines planes, points, or axis movement conditions in an assembly.

Have at least two parts on the Model screen that need to be assembled.

Type **AMMATE** ↵ (the "mate" assembly constraint option) or click the icon.

Select first set of geometry: (*Click the first part near the face to be mated. Continue to click the left mouse button until the face is highlighted with the arrow pointing out, then press ↵ or the right mouse button.*)

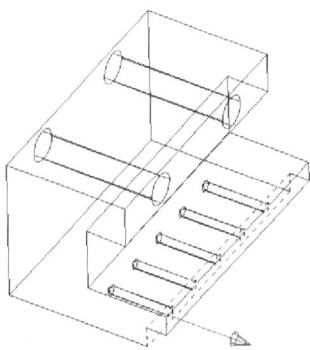

Select second set of geometry: *(Click the second part near the face to be mated. Continue to click the left mouse button until the face is highlighted with the arrow pointing out, then press ↵ or the right mouse button.)*

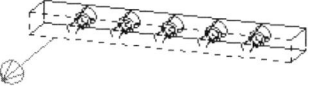

Enter Offset <0.000>: *(Since you want no gap between these two surfaces, press ↵ to accept the 0.000 offset.)*

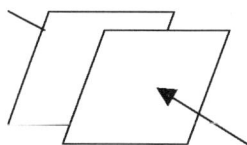

At this point you have only instructed the two faces to mate. You have not given any other constraint instructions, so the assembly may look somewhat awkward at this point.

You can move the parts away from each other to make it easier to see surfaces. This will have no effect on the constraints.

The next step is to add more constraints.

Step 5   Type **AMMATE** ↵

or click the icon.

Select first set of geometry: *(Select on the edge lines, click Return, and select the mating axis lines.)*

Select second set of geometry: *(Select on the edge lines, click Return, and select the mating axis lines.)*

(Note: Make sure the axis lines run in the same direction.*)*

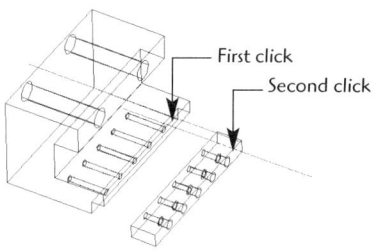

Enter Offset: <0.000> *(Press ↵ to accept the 0.000 offset.)*

You need one more constraint to keep the parts from spinning around the first axis line.

Step 6 Type **AMMATE** ↵
or click the icon.

Select first set of geometry: *(Select on the edge lines, press ↵, and select the mating axis lines.)*

Select second set of geometry: *(Select on the edge lines, press ↵, and select the mating axis lines.)*

(*Note:* Make sure the axis lines run in the same direction.)

Enter Offset: <0.000> *(Press ↵ to accept the 0.000 offset.)*

The two parts are now well constrained as an assembly.

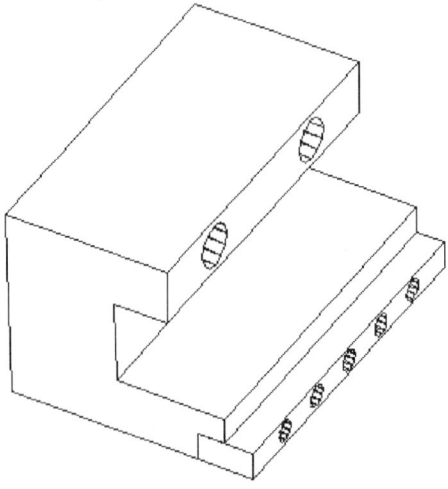

ASSEMBLIES | **179**

The next step is to use an assembly constraint called *insert* after bringing in more parts from the catalog.

Step 7 Type **AMCATALOG** ↵
or click the icon.

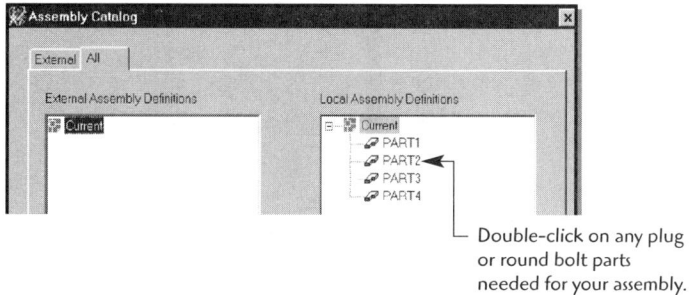

Double-click on any plug or round bolt parts needed for your assembly.

Right-click, then click **OK** when you have finished inserting from the catalog.

Step 8 Zoom into the area to perform the insert.

Type **AMINSERT** ↵
or click the icon.

`Select first circular edge:` *(Select the inside circular edge.)*

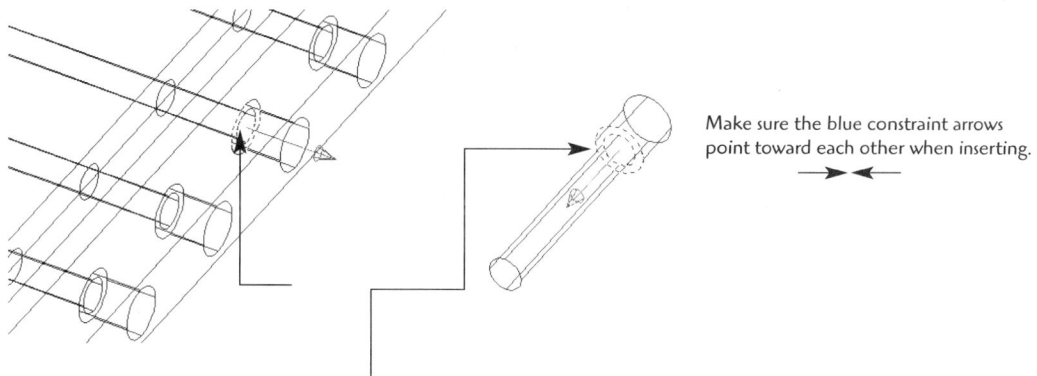

Make sure the blue constraint arrows point toward each other when inserting.
→ ←

`Select second circular edge:` *(Select the inside circular edge.)*
`Enter offset <.0000>:` ↵

The insert is complete.

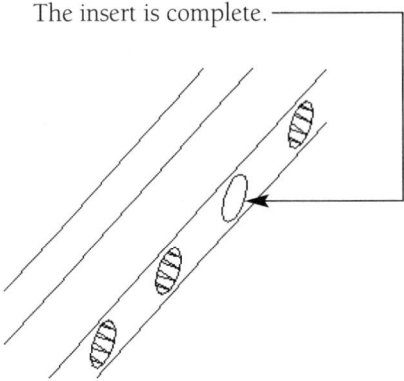

Continue inserting and mating parts until all assembly parts are constrained.

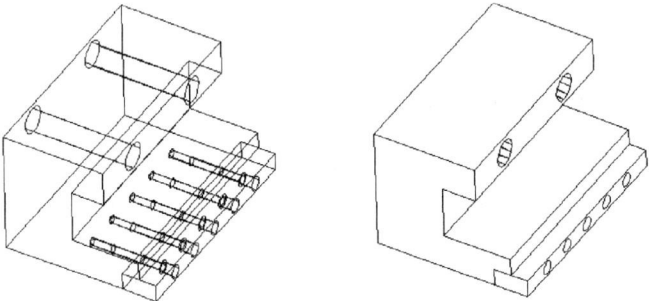

You are now ready to create the assembly view or exploded view, which is done by creating a **scene**.

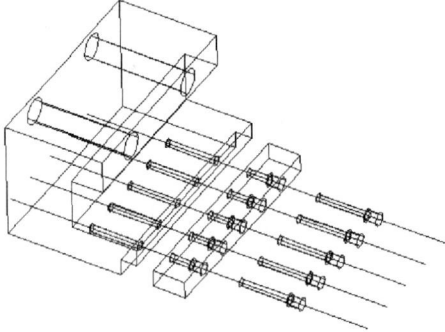

ASSEMBLIES | *181*

Step 9 Type **AMNEW** ↵
or click the icon.

```
Specify target or assembly name <DRAWING>: ↵
Enter new scene name of the active assembly <SCENE1>:
```
(Type in a scene name.)

```
Enter overall explosion factor <.0000>: 5
```
(Type in an explosion factor to separate the parts.)

Click **OK**.

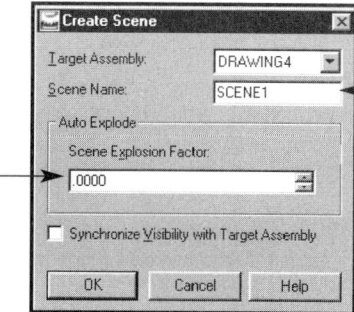

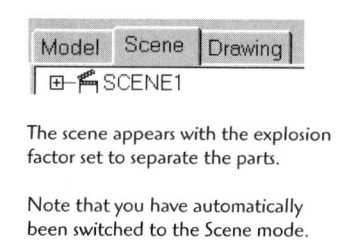

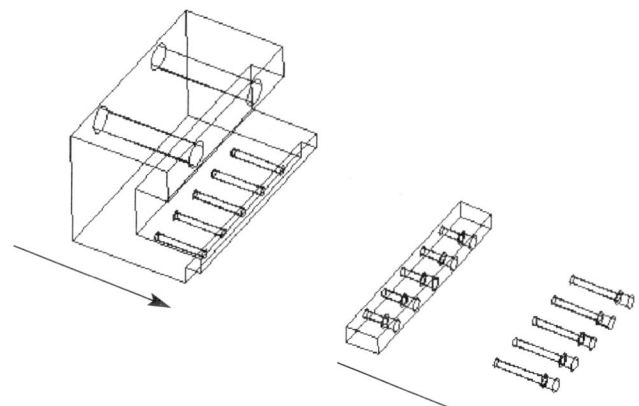

Let's add some details to improve the exploded assembly view.

Step 10 To add trails between parts
type **AMTRAIL** ↵
or click the icon.

```
Select reference point on part or subassembly:
```
(Click a center point on the assembly.)

CHAPTER 11

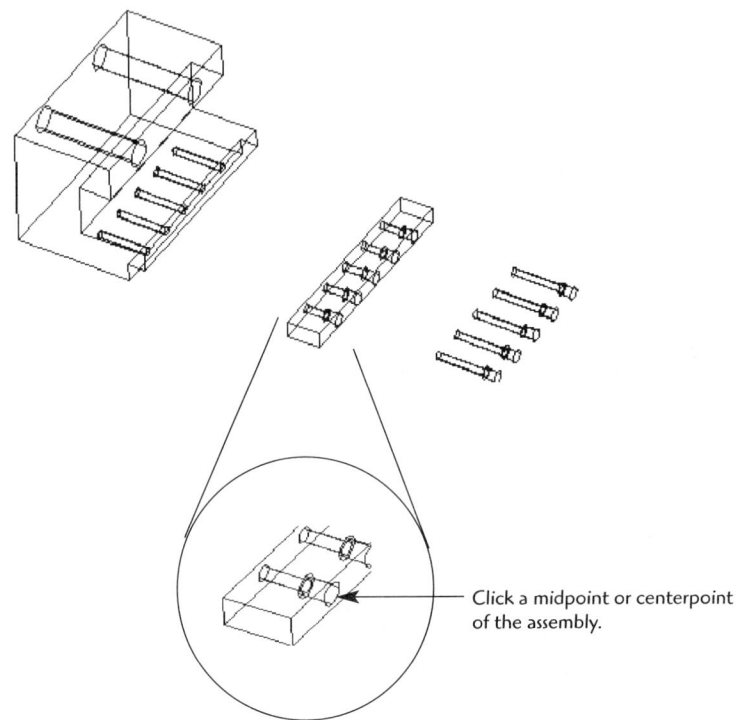

Click a midpoint or centerpoint of the assembly.

The Trail Offsets dialog box appears.
Enter positional offsets. (Higher values will overshoot the trail a farther distance. Using the Pick option can help if the trail points are not obvious.)
Click **OK**.

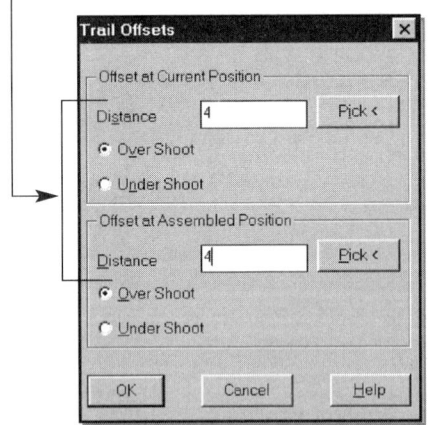

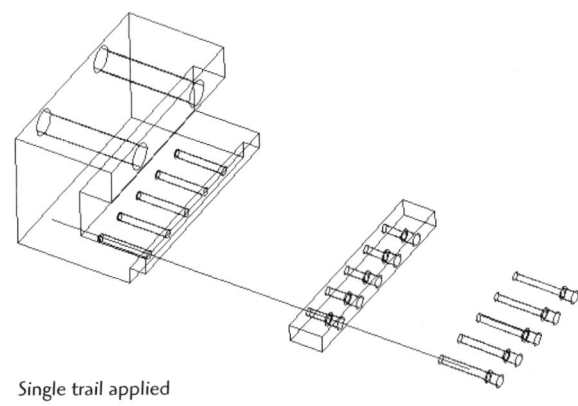

Single trail applied

Continue adding trails to all insertion points of the assembly.

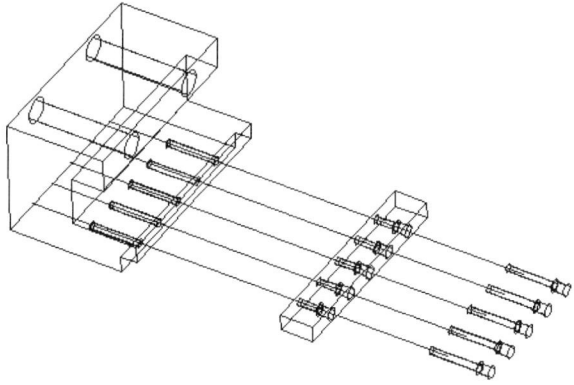

The next step is to add these exploded assemblies to the orthographic projections.

Step 11 Type **AMDWGVIEW** ↵

or click the icon.

The Create Drawing View dialog box appears.

Keep the View Type set as **Base**.

The Data Set is set to the Scene view.

Reduce the Scale initially so that the view fits into the drawing mode scene. You can increase the scale later if necessary.

Click **OK**.

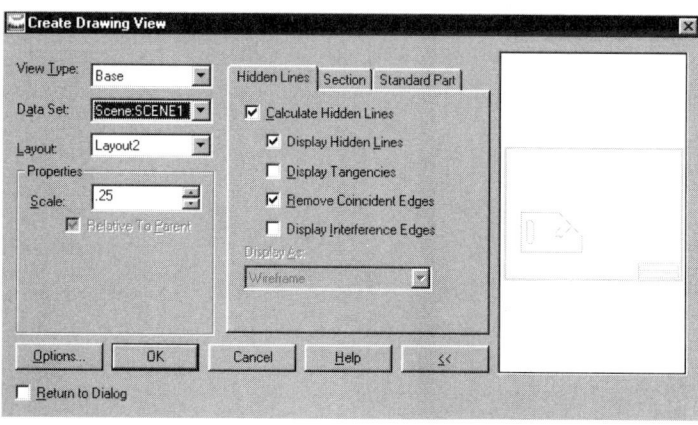

```
Select planar face, work plane or [Ucs/View/worldXy/
worldYz/worldZx]: V ↵↵ (Type V for view, then Enter, Enter.)
```

Click to place the view on the drawing page.

You may add other views of the assembly.

These views are base views with the front selected as the alignment. Then, add ortho views of the top and side.

You can create other combinations of views of the assembly and the individual details of each part, but first you need to activate the individual parts.

Type **AMACTIVATE** ⏎.

`Enter an option [Assembly/Part/Scene] <Part>:` *(Type P for part and ⏎.)*

`Select part to activate or [?] <PART1_1>:` *(Click on one of the parts for which you will be creating an orthographic drawing.)*

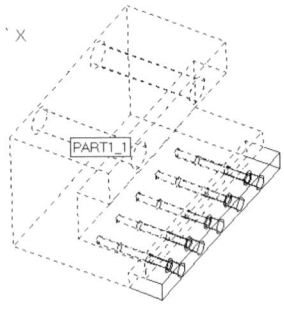

Now that the part is activated, you will create the orthographic drawing views just as you learned in Chapter 8.

Type **AMDWGVIEW** ⏎
or click the icon.

The Create Drawing View dialog box appears.

Make sure **Active Part** is selected in the Data Set. Start with the base view and then add orthographic views.

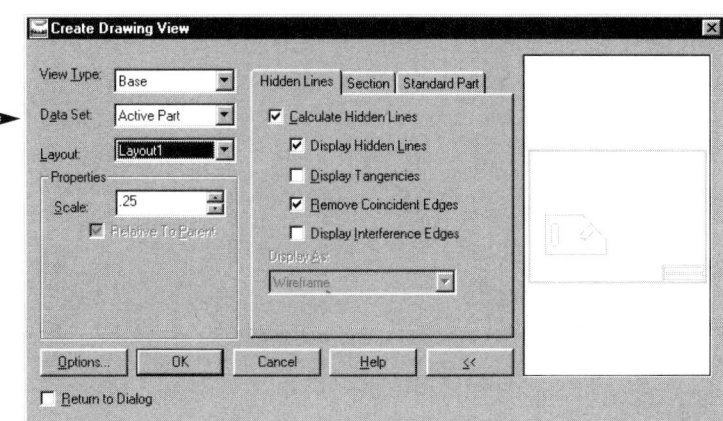

Activate the other parts with the AMACTIVATE command, and continue to create orthographic views.

The combination of the exploded assembly views and the orthographics of each part produces a high-quality drawing.

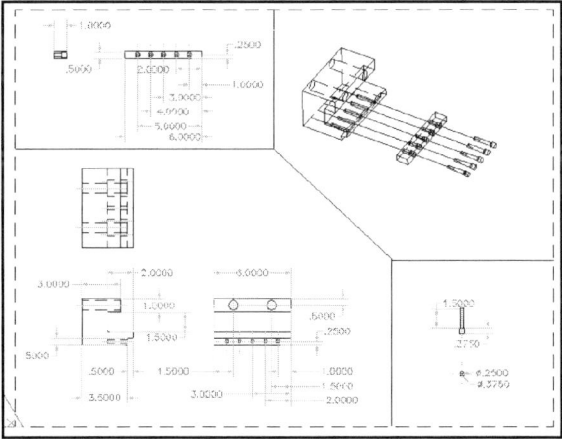

For a Deeper Understanding about
Assemblies

Understanding the Assembly Constraining Process

There are four primary types of assembly constraints:

- Mate
- Flush
- Insert
- Align

Mate is for mating planes, lines, and points. There are three Mate options:

- **Plane** (forces planes to face each other)

- **Lines** (Places lines in-line)

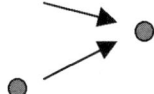

Two separate lines or edges

Now the lines are constrained in-line.

Point (attaches points)

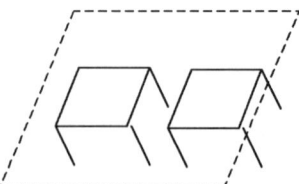

Flush is for placing surfaces in the same plane (forces surfaces to *stay* in the same planes).

Insert is for placing plugs, pins, bolts, and screws into holes. (*Note:* The blue arrows should always face each other.)

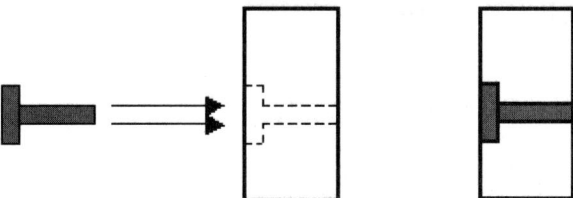

Align is for placing surfaces at an angle to one another (forces planes or edges at an angle).

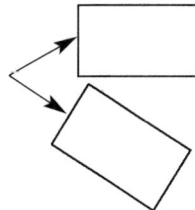

Editing the Assembly Constraints

The desktop browser is a powerful editing tool for editing the assembly constraints. You will see the constraint labels in the browser.

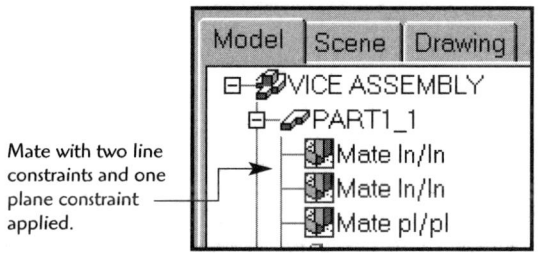

Mate with two line constraints and one plane constraint applied.

You can edit these assembly constraints by right-clicking on them in the desktop browser.

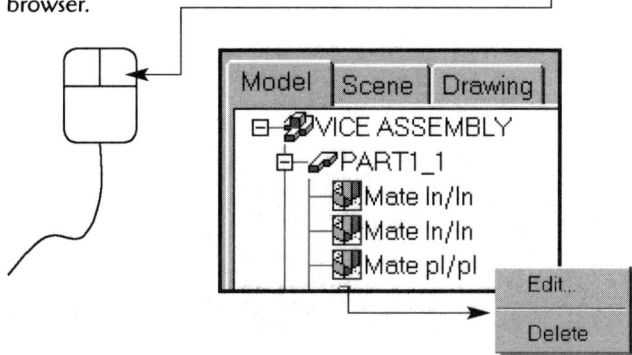

Two options appear, **Edit** and **Delete.**

- Deleting will remove the constraint from the part. *(This is helpful for reapplying a constraint that did not produce the expected results.)*
- Edit will bring up an edit box.

The Expression option sets the distance between faces, lines, or points, depending on constraint type. The Expression option can be toggled, and the assembly will update dynamically.

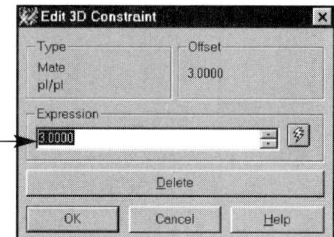

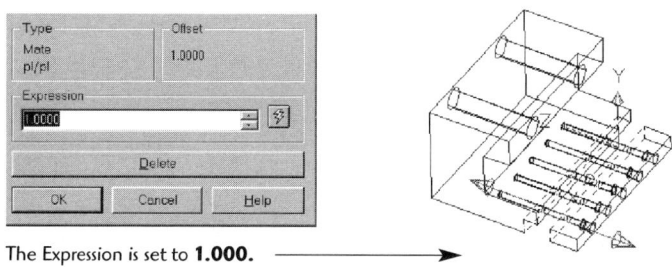

The Expression is set to **1.000.**
Note the part spacing.

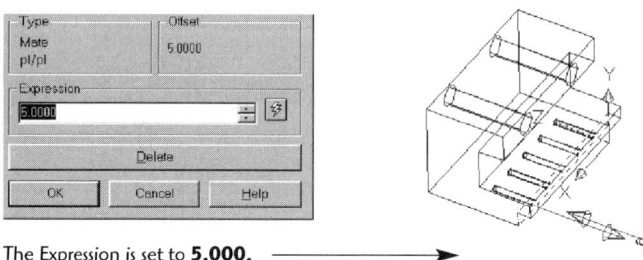

The Expression is set to **5.000.**
Note the part spacing.

Note also that when you pass the cursor over the constraint in the desktop browser, the blue constraint arrows appear to indicate how the constraint is applied.

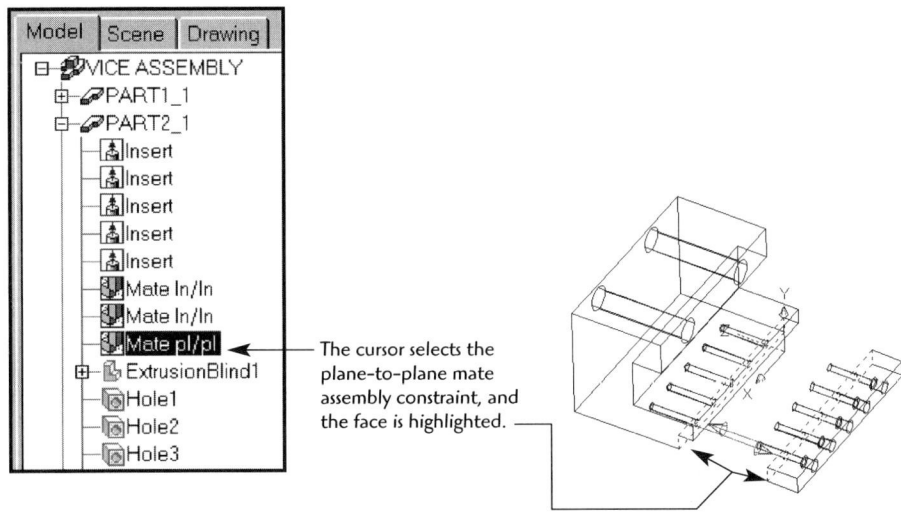

The cursor selects the plane-to-plane mate assembly constraint, and the face is highlighted.

ASSEMBLIES AS DESIGN TOOLS

Assemblies are useful design tools that can give important design information about the design. We will first look at examining the **mass properties** of the assembly. These properties include:

- **Mass** (weight)
- **Volume** (cubic area)
- **Center of gravity** (center balance point)
- **Inertia** (stiffness on axes)
- **Moments** (load times distances on axes)
- **Radius of gyration** (radial movement of compressional load on an axis)

(*Note:* These are somewhat simplified descriptions of these properties, but they should help the nonengineer gain some understanding of the properties.)

Checking the Mass Properties of an Assembly

Step 1   Type **AMMASSMPROP** ↵.
(*Note:* Use **AMASSMPROP** for older versions of MDT.)
or click the icon.

    Select part and subassembly instances: *(Use a **crossing window** to select the entire assembly, or the parts of the assembly that you need.)*

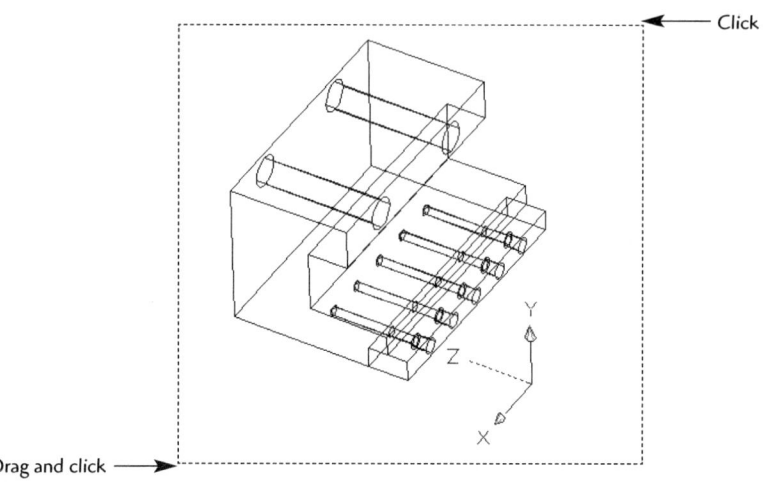

Drag and click

Click

Step 2 The first Assembly Mass Properties dialog box appears. Here you can set the **material** for the parts, the **units of measure**, and the **coordinate system location** from which properties are measured.

Step 3 Select the materials from which your assembly parts are made. The display shows each material's properties. Each part in the assembly can be assigned a material.

Click **Assign Material**. (A dialog box will ask you to confirm the assignment; click **OK**.)

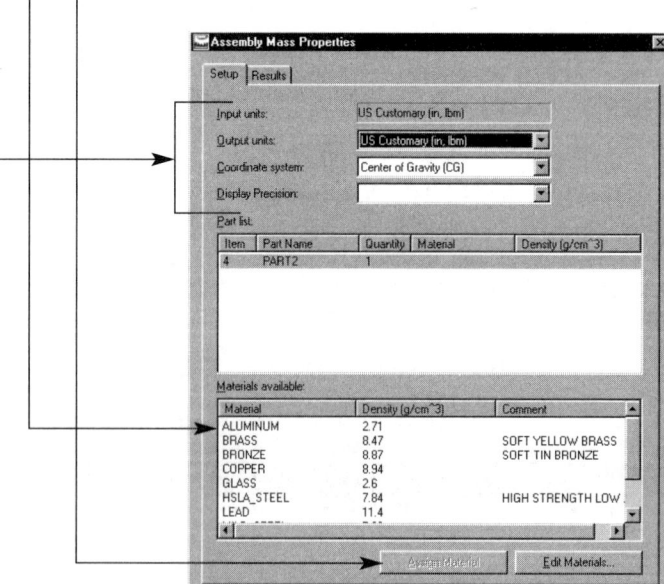

ASSEMBLIES | 191

Step 4 Click on the **Results** tab, then click **Calculate**. The mass properties data appear. By clicking **Export Results**, you can create and send a text file of the data to any location. This file can be sent and attached to the drawing. The file can be recognized by the extension .MPR, for mass properties.

To import the data into the drawing, use the MTEXT command, drag a location window, and use **Import File**. Set the file name to *.MPR and navigate to your file. Set any other text settings necessary and click **OK**.

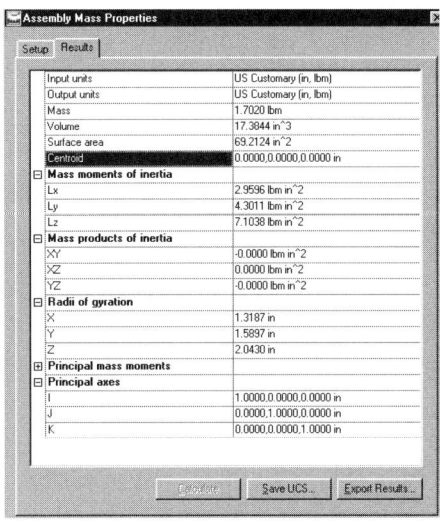

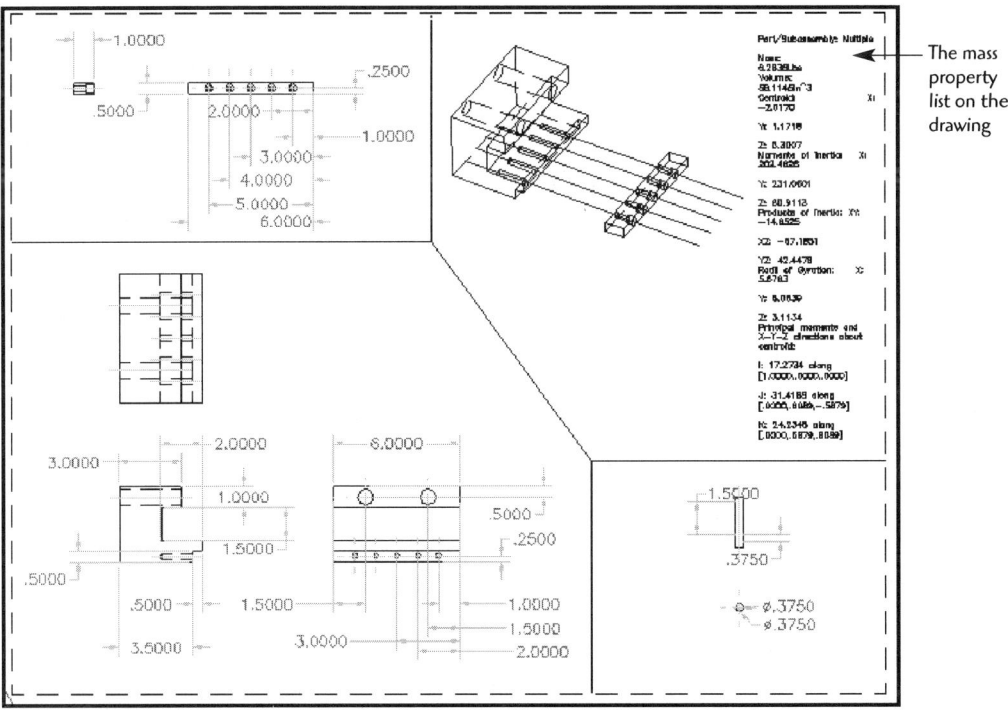

The mass property list on the drawing

Checking Interference on an Assembly

Another helpful assembly designing tool provided in Mechanical Desktop is the ability to check part interference. Sizes can then be corrected in the assembly.

Step 1   Type **AMINTERFERE** ↵
or click the icon.

`Nested part or subassembly selection? [Yes/No] <No>:`
(Select Enter to confirm "No." Only select yes if you have nested parts inside your main part.)

Step 2   `Select first set of parts or subassemblies:` *(Select one of the parts to be tested, then ↵.)*

`Select second set of parts or subassemblies:` *(Select one of the parts to be tested, then ↵.)*

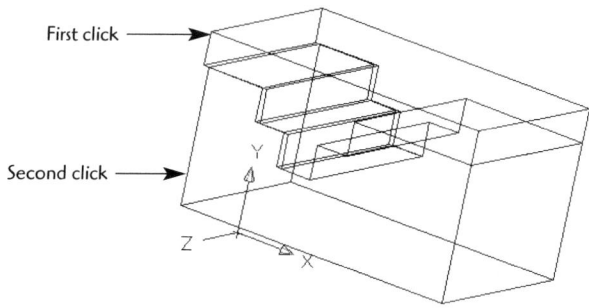

First click
Second click

Step 3   `Parts/subassemblies do not interfere.` *(This message will appear in the command line **only** if parts **do not** interfere.)*

`Create interference solids? [Yes/No] <No>:` y ↵ *(This message will appear in the command line **only** if parts **do** interfere. Creating solids from the interference can help edit out the problem areas of the assembly.)*

`Highlight pairs of interfering parts/subassemblies? [Yes/No] <n>:` ↵ *(Answering Yes will highlight the parts that touch or interfere with each other.)*

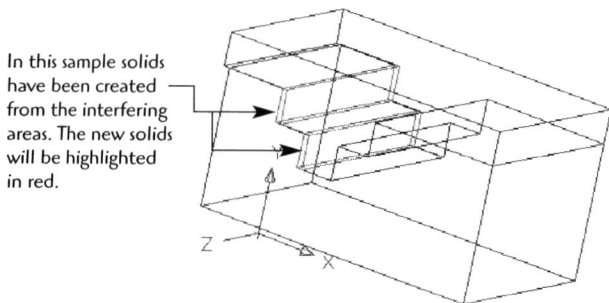

In this sample solids have been created from the interfering areas. The new solids will be highlighted in red.

Measuring Distances with the 3D Distance Command

Step 1   Type **AMDIST** ↵
or click the icon.

Step 2   `Select first set or [Objects]:` *(Select one of the parts to be measured, then ↵.)*
`Select second set or [Objects]:` *(Select one of the parts to be measured, then ↵.)*

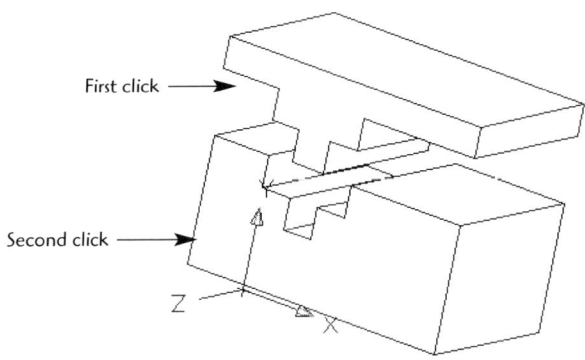

Step 3   `Enter output type [Display/Line] <Display>:` ↵
`Minimum distance:<0.853031>`

Drawing Project 11–1

Assembly Drawing Development

For this project you will produce an assembly drawing similar to the following drawing. You should model all the component parts first. Remember to use the AMNEW command before creating each part so it is placed in the catalog. Create a new exploded view scene for the assembly. Finally, place the exploded view of the assembly along with orthographic views of each part into the final drawing mode view. The drawing should be detailed enough that a reader could produce the part. When you have finished, print out a copy of the drawing for your instructor.

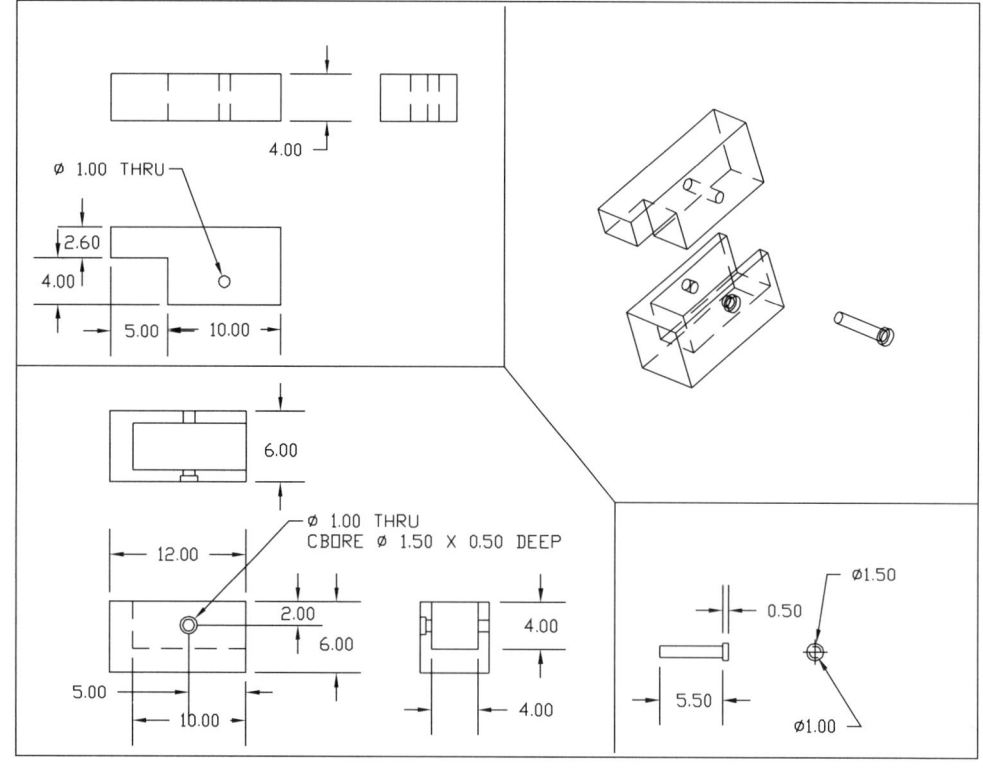

ADVANCED ASSEMBLIES AND BILL OF MATERIALS

The **Assembly Modeling** toolbar offers shortcuts for most of the assembly-related commands.

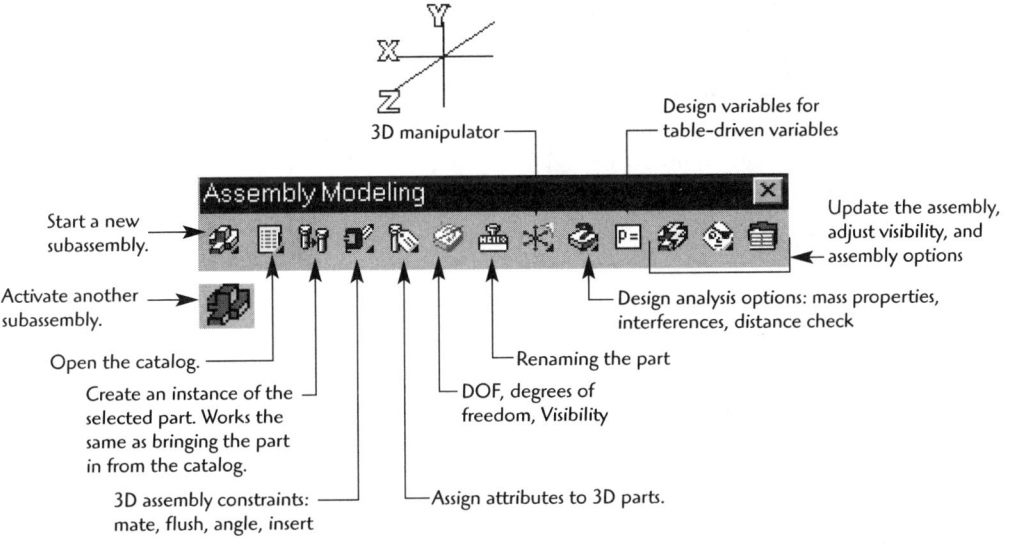

Using the 3D Manipulator to View the Assembly

Type **AMMANIPULATE** ↵
or click the icon.

(Click and Drag on an X-, Y-, or Z-axis to manipulate the assembly component.)

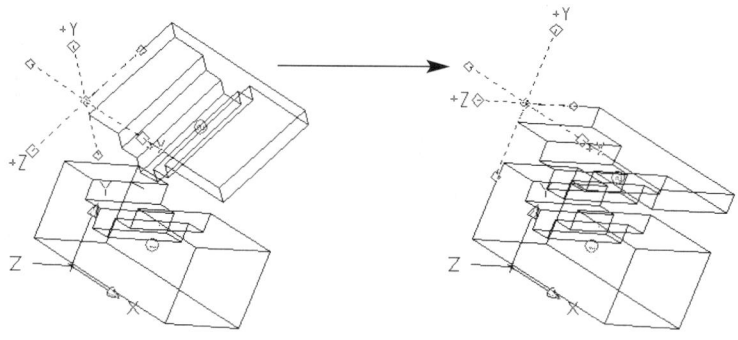

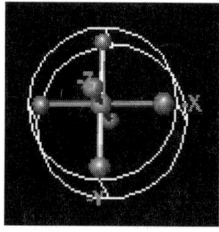
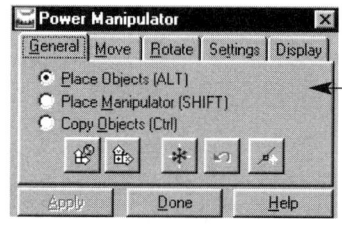

Parts of the assembly can be moved and rotated dynamically or with keyed distances with the dialog box.

Developing a Bill of Materials (BOM)

As you create your assembly drawing, Mechanical Desktop keeps track of the drawing data. This allows for an automated development of a **bill of materials** (BOM).

Try these steps to add a bill of materials to one of your assembly drawings.

Type **AMBOM** ↵
or click the icon.

The BOM (Bill of Materials) dialog box appears.
The box lists:

- Item number
- Quantity
- Part name
- Material type
- Part notes
- Part vendor

ADVANCED ASSEMBLIES AND BILL OF MATERIALS | *197*

Double-click in any of the boxes to edit the values.

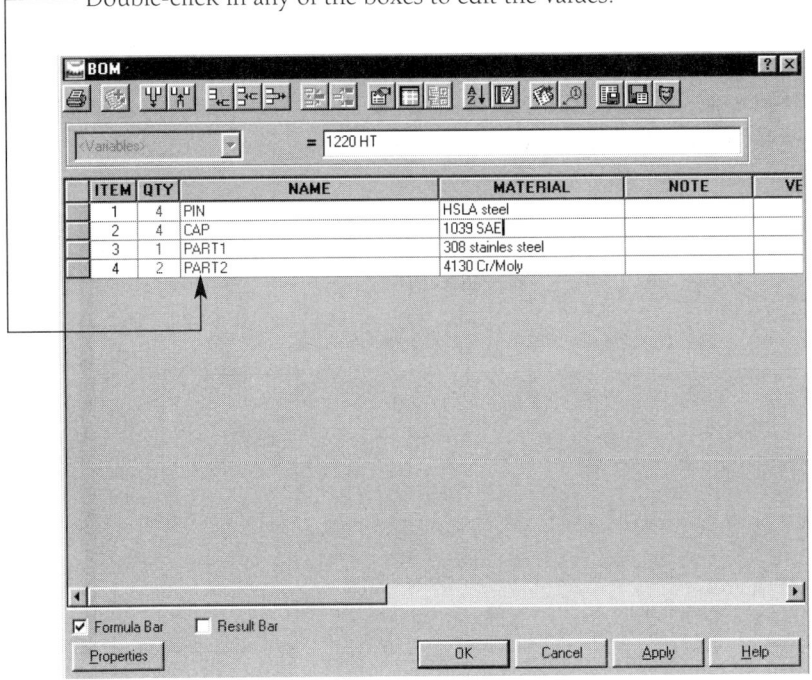

Other features can be edited on the toolbar.

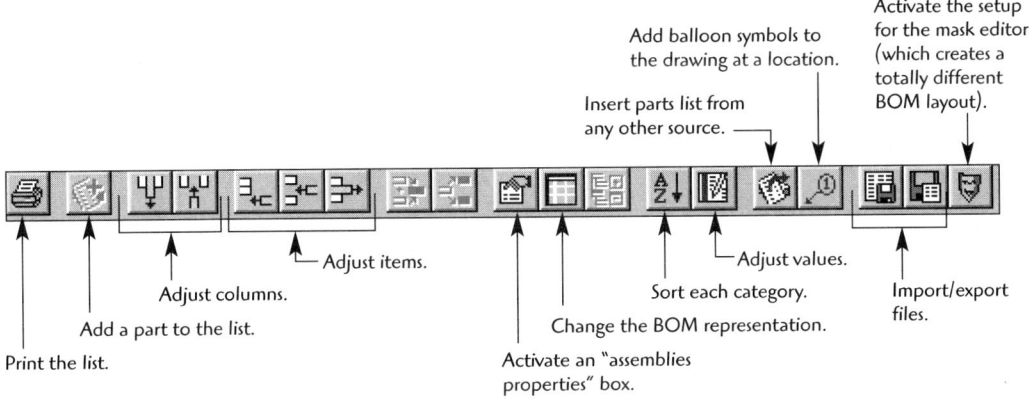

Once you have made any adjustments to your bill of materials, you are then ready to save the BOM file.

Click the Export button.
The Export dialog box appears. Set the file type to "*.txt." (*Note:* Other export file types can be used.)

Type in the file name.

Click **OK**.

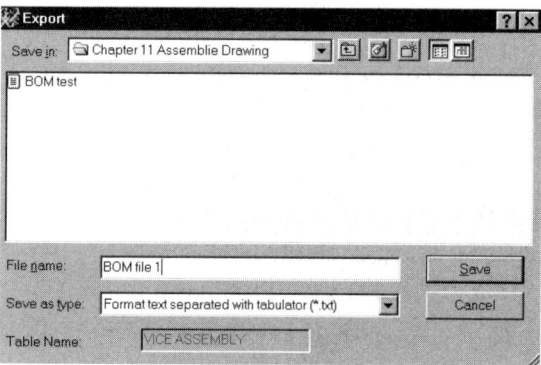

Placing the Bill Of Materials on the Drawing

Once you have saved your bill of materials, you are then ready to place it into the drawing. The basic process is the same as for placing any text document into a drawing using the MTEXT command.

Click into the **Drawing** mode tab in the desktop browser.

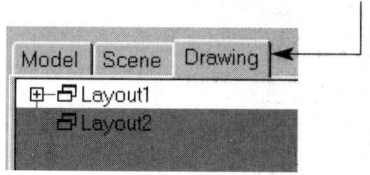

Type **MTEXT** ↵
or click the icon.

Click and drag a window for the BOM text location.

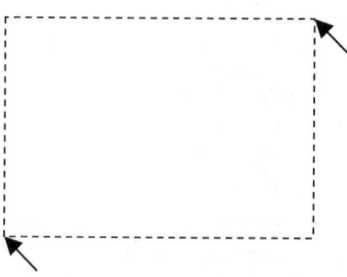

When the Multiline Text Editor appears, click on **Import Text** and navigate to and select the BOM file.

Click **OK**.

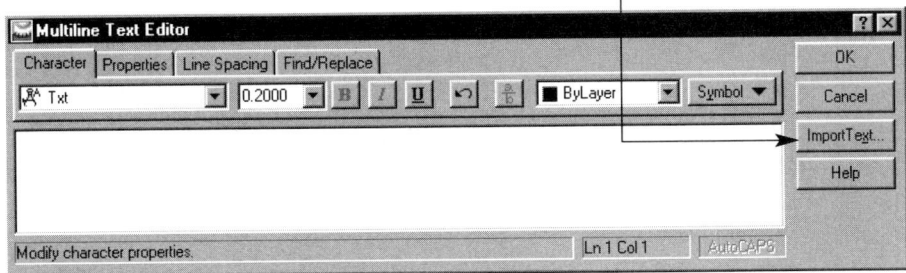

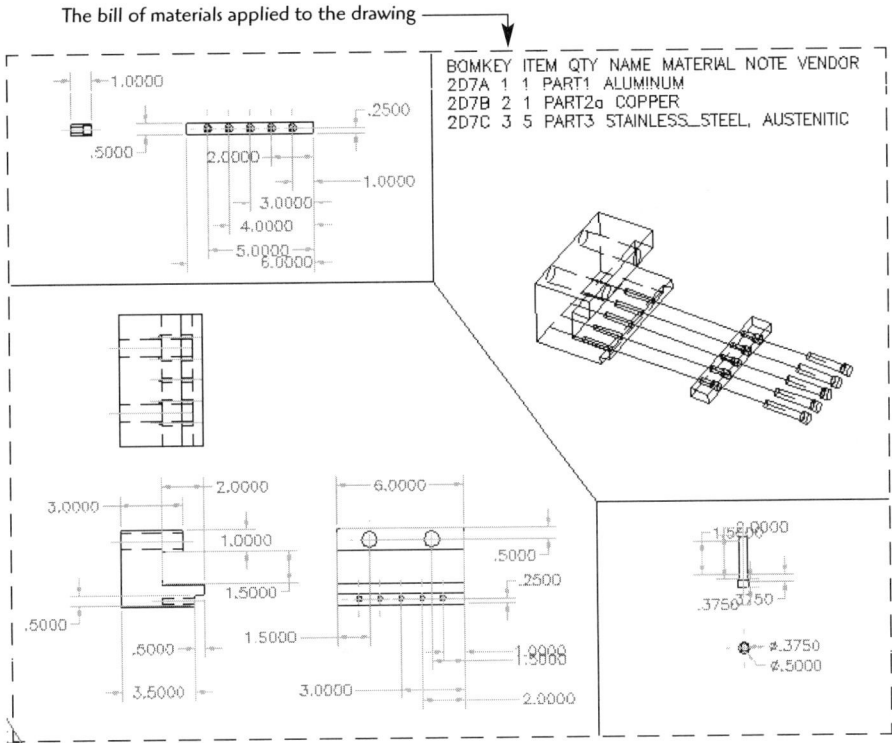

The bill of materials applied to the drawing

The bill of materials can also be placed directly into the drawing.

Type **AMPARTLIST** ↵.

The Parts List dialog box appears. (Make any adjustments or additions to the list as required.)

Click **OK**.

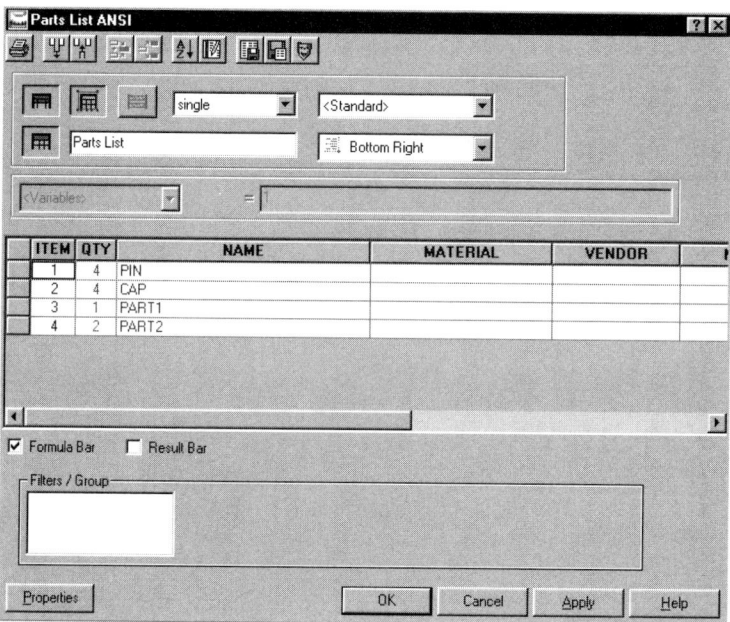

`Specify location:` *(Move the bill of materials into its location.)*

(Note: You can use the Scale command to adjust the size of the bill of materials box.)

ADVANCED ASSEMBLIES AND BILL OF MATERIALS | 201

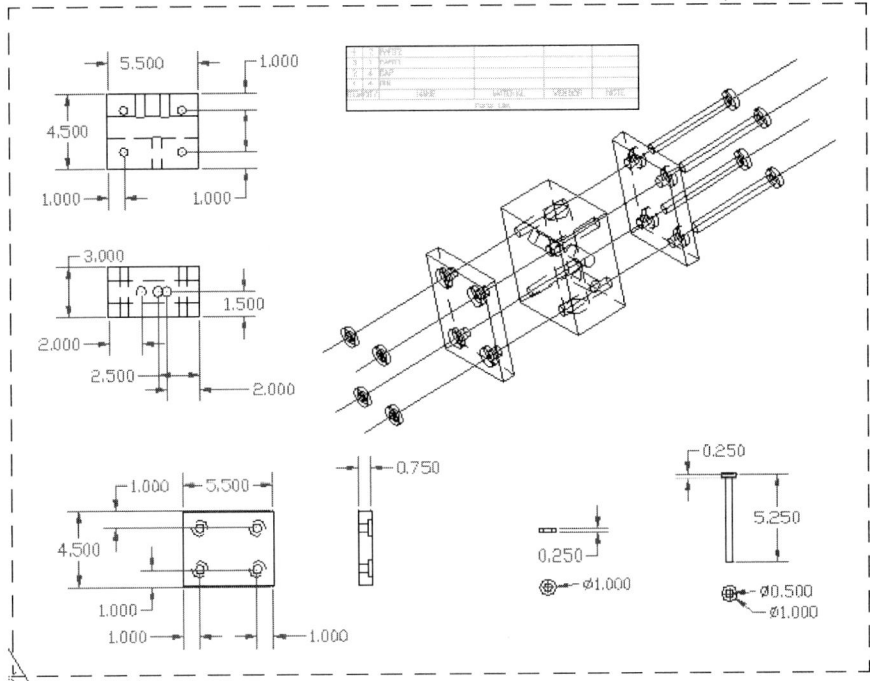

The completed assembly drawing with bill of materials

To change the color of the text in the material list:

Type **AMEDIT** ↵.

Pick the material list box.
The Parts List dialog box appears.

Click into the text area, then click the **Properties** button.

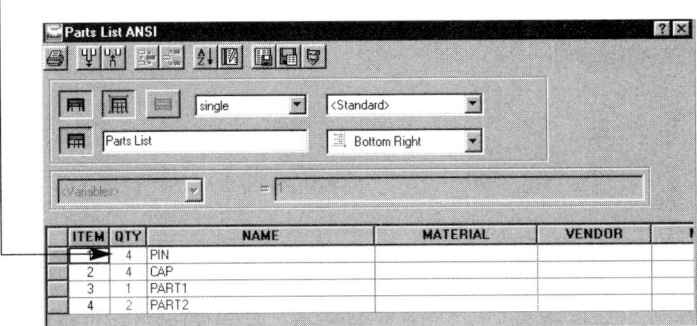

The Parts List Properties for ANSI dialog box appears, where you can make many adjustments to the parts list.

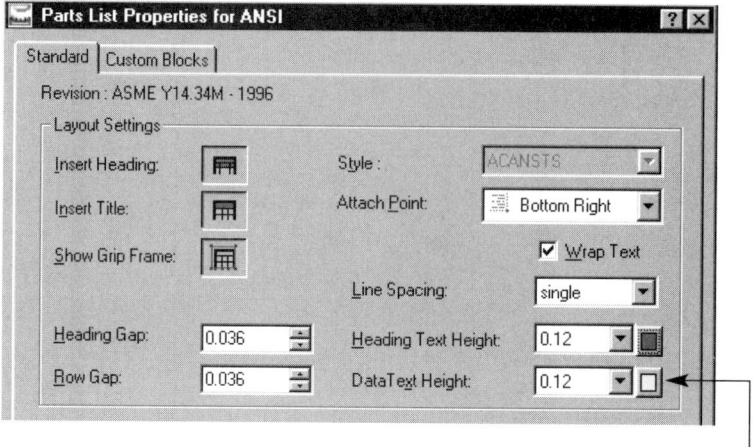

To change text color, click the color box and select a new color.

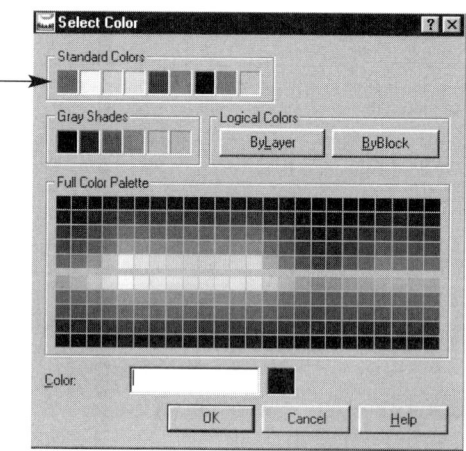

Once you have selected the color, click **OK, OK,** and **OK** again to return to the drawing screen.

Attaching Balloon Leaders

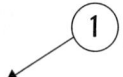

Type AMBALLOON ↵.

```
Select part/assembly:  (Click on one of the parts.)
Select next point:     (Drag and click the balloon into position.)
```

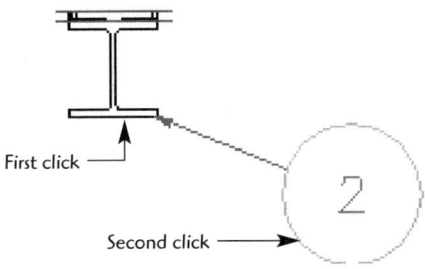

A completed drawing with a bill of materials and balloon leaders

Note how the numbers in the bill of materials match up with the balloon leaders.

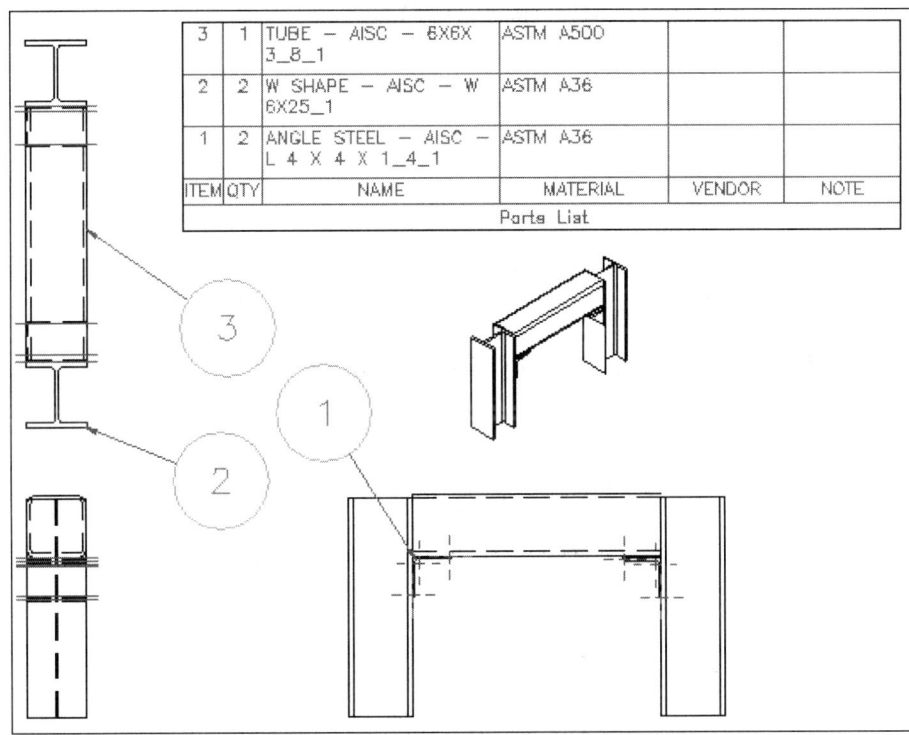

3	1	TUBE – AISC – 6X6X 3_8_1	ASTM A500		
2	2	W SHAPE – AISC – W 6X25_1	ASTM A36		
1	2	ANGLE STEEL – AISC – L 4 X 4 X 1_4_1	ASTM A36		
ITEM	QTY	NAME	MATERIAL	VENDOR	NOTE
		Parts List			

Drawing Project 12-1

Assembly Drawing Development and Bill of Materials

For this project you will again produce an assembly drawing similar to the following drawing, including the bill of materials. You should model all the component parts first. Remember to use the AMNEW command before creating each part so it is placed into the catalog. Create a new exploded view scene and bill of materials for the assembly. Finally, place the exploded view of the assembly along with orthographic views of each part into the final drawing mode view. The drawing should be detailed enough that a reader could produce the part. When you have finished, print out a copy of the drawing for your instructor.

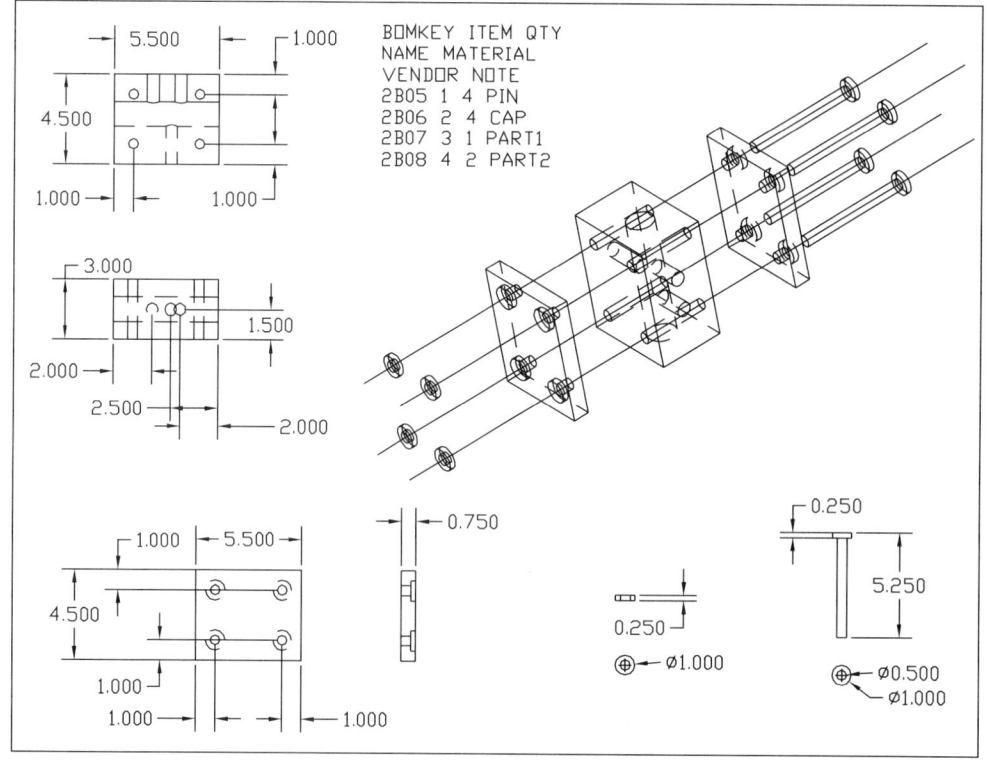

Drawing Project 12-2

Assembly Drawing Development: Pulley

For this project you will produce an assembly drawing of the pulley from the following data. You should model all the component parts first. Remember to use the AMNEW command before creating each part so it is placed into the catalog. Create a new exploded view scene and bill of materials for the assembly. Finally, place the exploded view of the assembly along with orthographic views of each part into the final drawing mode view. The drawing should be detailed enough that a reader could produce the part. When you have finished, print out a copy of the drawing for your instructor.

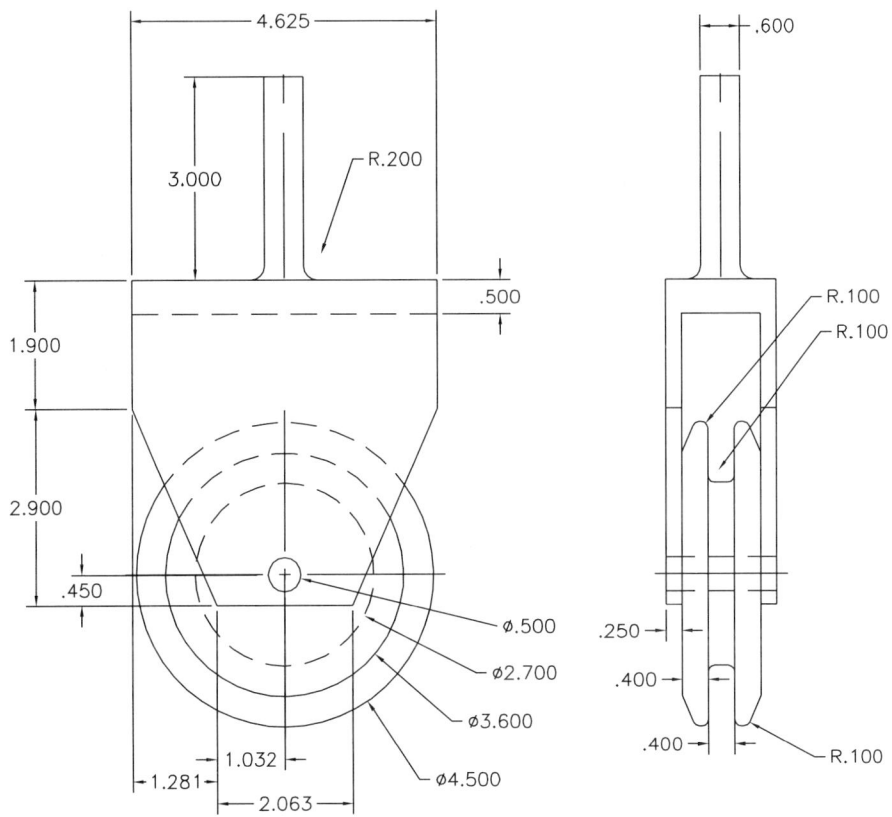

ADDING SYMBOLS TO A DRAWING

Drawings contain symbols to convey the intent of the engineer. The use of correct symbols improves the accuracy and quality of the final product. Additionally, symbols provide a common language among engineers, fabricators, contractors, and manufacturers. Mechanical Desktop provides a ready list of commonly used symbols. We will look at the following symbols that Mechanical Desktop provides:

- Surfacing/machining
- Welding
- Geometric tolerances

Applying Surfacing/Machining Symbols

Open a drawing that requires symbols.

Type **AMSURFSYM** ↵
or click the icon.

`Select object to attach:` *(Click on the surface.)*

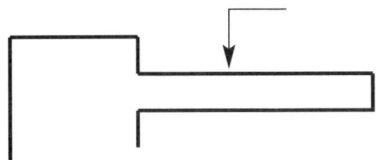

Start Point: *(Click where the arrowhead should be placed.)*
Next Point <Symbol>: *(Drag the arrow and reference line, then ↵.)*

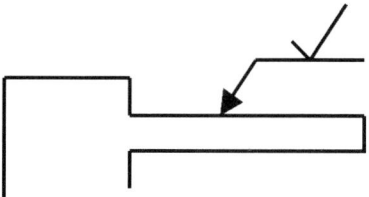

The Surface Texture dialog box appears.

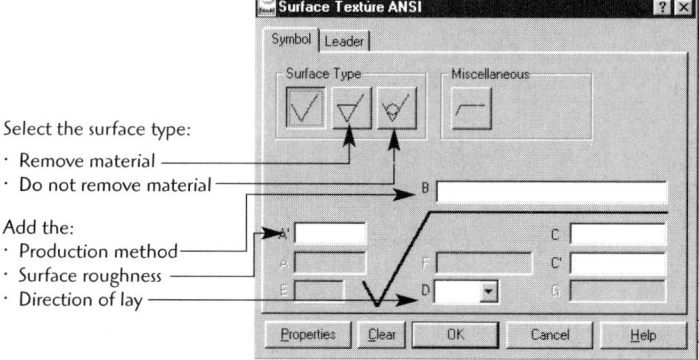

Select the surface type:
- Remove material
- Do not remove material

Add the:
- Production method
- Surface roughness
- Direction of lay

Click **OK**.

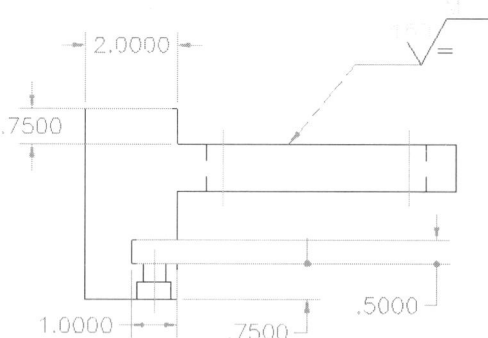

The surfacing symbol is applied to the orthographic view.

Adding Symbols to a Drawing | 209

Applying Welding Symbols

Open a drawing that requires symbols.

Type **AMWELDSYM** ↵
or click the icon.

`Select object to attach:` *(Click in the part.)*

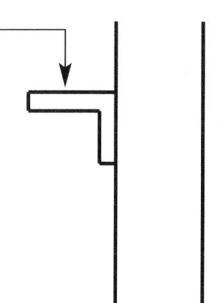

`Start Point:` *(Click where the arrowhead should be located.)*
`Next Point: <Ortho off>` *(Drag the arrow and reference line, then ↵.)*

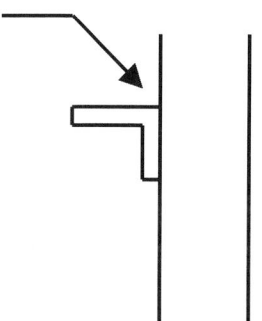

The Weld Symbol dialog box appears.

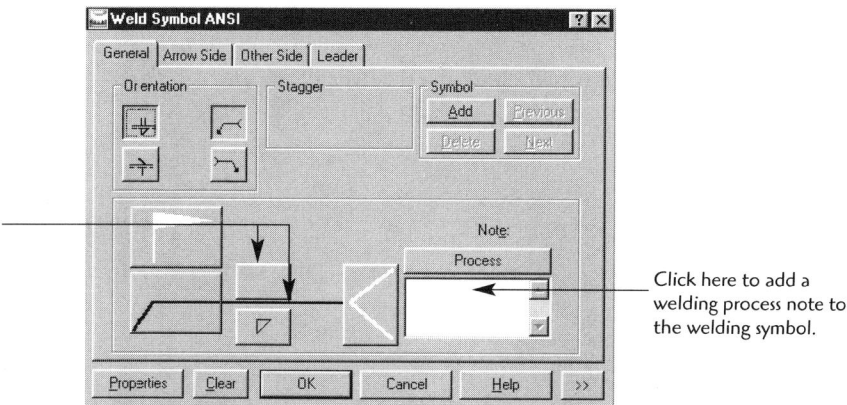

Click here to add a welding process note to the welding symbol.

Add the type of symbol by clicking the Symbol button. The list will be displayed. Click on a weld type.

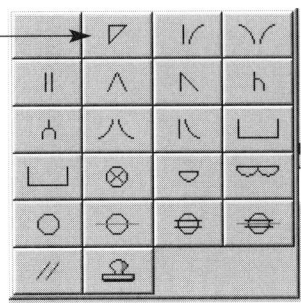

Each type of weld symbol requires dimensional properties. The *fillet* weld in this example requires leg sizes, and a length and pitch. Additionally, the contour and method of finishing symbol can be added.

Enter the properties and click **OK**.

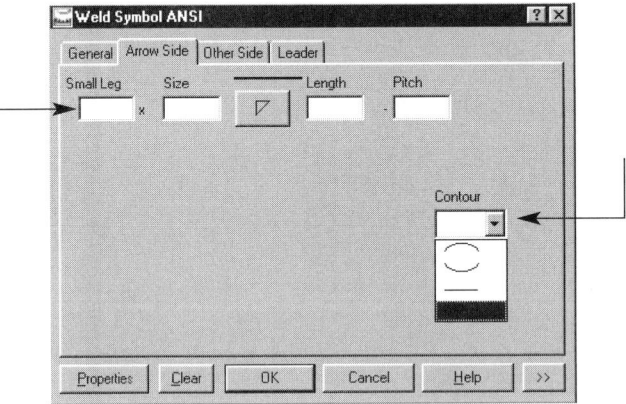

The final weld symbol is shown with the following properties:

- 1/4" weld leg size
- 2" length of weld
- 6" pitch (center-to-center)
- flush by grinding

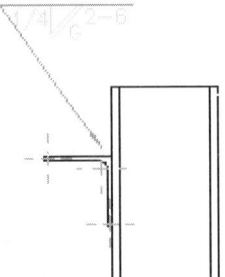

The following is a list of typical welds and associated welding symbols:

▷	Fillet Weld
⼁\	Flange Corner Weld
⼂\	Flange Edge Weld
‖	Square Groove Weld
V	V Groove Weld
⼃/	Bevel Groove Weld
⼄	J Groove Weld
Y	U Groove Weld
⼅	Flare V Groove Weld
⼆	Flare Bevel Groove Weld
⊓	Plug Weld
⊓	Slot Weld
⊗	Stud Weld
⌒	Back or Backing Weld
∽∽	Surfacing
○	Spot or Projection Weld
⊖	Spot Weld on Reference Line
⊖	Seam Weld

Applying Geometric Tolerance Symbols

Open a drawing that requires symbols.

Type **AMFCFRAME** ↵
or click the icon.

`Select object to attach:` *(Click in the part.)*

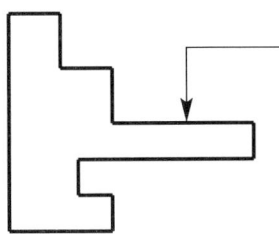

`Start Point:` *(Click where the leader line should be located.)*
`Next Point: <Ortho off>` *(Drag the leader line, then ↵.)*

The Feature Control Frame dialog box appears.
Select the symbols and tolerances.

Click **OK**.

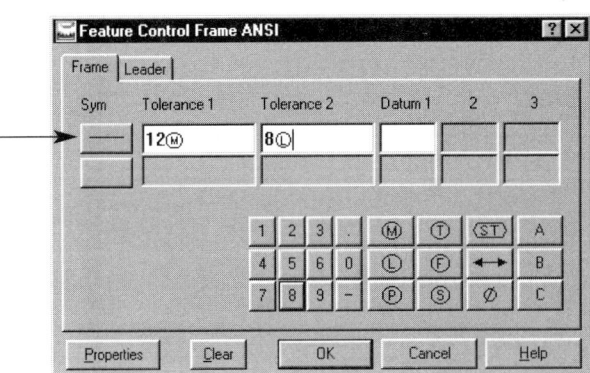

The geometric tolerances are applied.

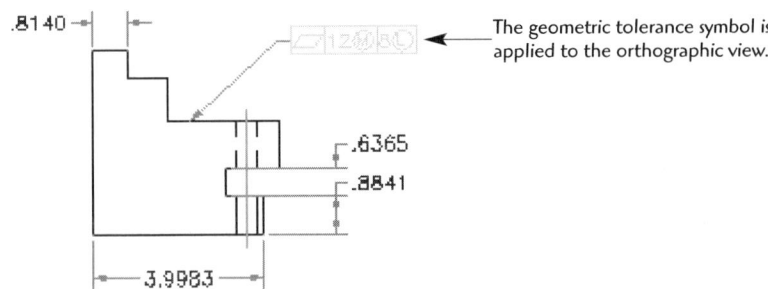

The geometric tolerance symbol is applied to the orthographic view.

ADDING SYMBOLS TO A DRAWING | *213*

Other variations of the geometric tolerancing symbol applications are available. These symbols are applied in the same way as those of the feature control system. Note the following icon applications.

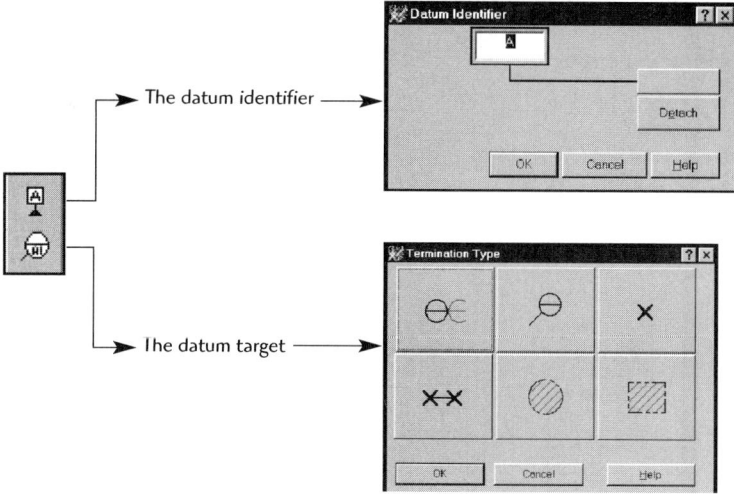

- The datum identifier
- The datum target

Drawing Project 13-1

Drawings with ANSI Symbols

Model the part and create an orthographic detailed drawing. Add the correct ANSI surface finish symbols as per the descriptions on the print. The drawing should be detailed enough that a reader could produce the part. When you have finished, print out a copy of the drawing for your instructor.

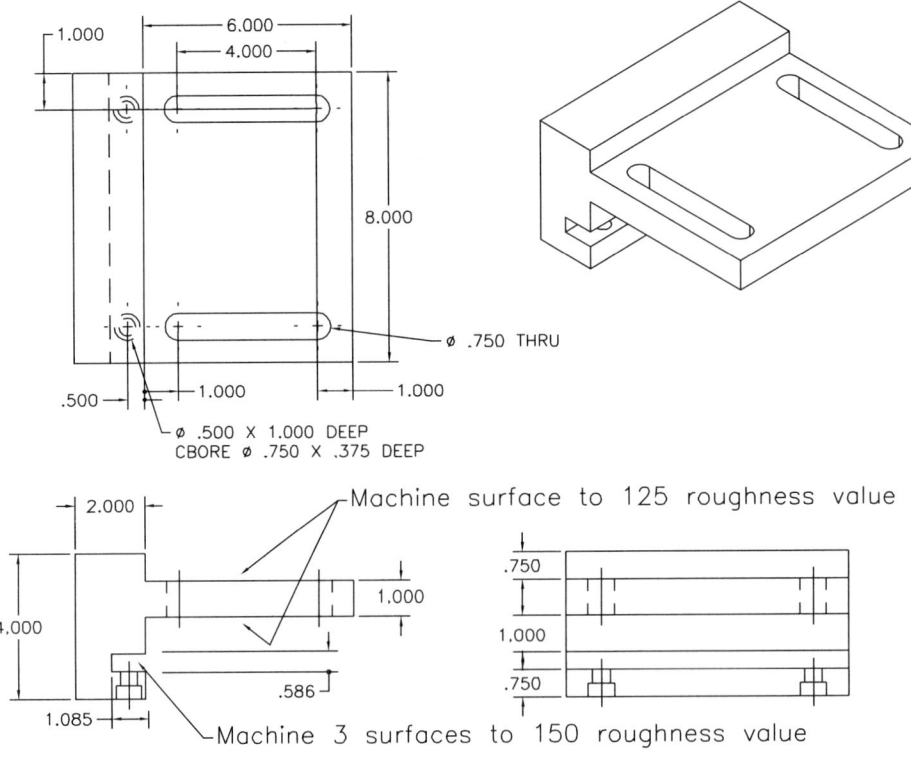

Drawing Project 13-2

Drawings with ANSI Welding Symbols

Model the part and create an orthographic detailed drawing. Add the correct ANSI welding symbols as per the descriptions on the print. The drawing should be detailed enough that a reader could produce the part. Chapter 14 will cover the creation of structural shapes from the power pack. So, you can create the angle and chanal shapes with the stated dimensions or review Chapter 14 before completing this drawing project. When you have finished, print out a copy of the drawing for your instructor.

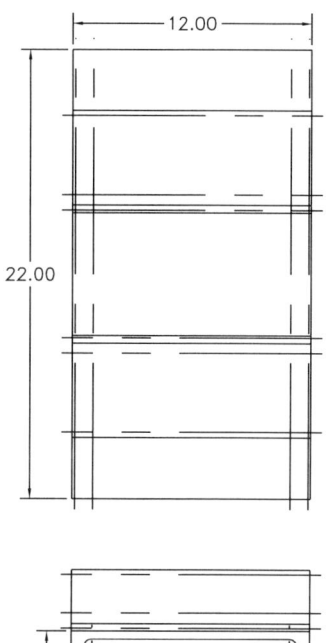

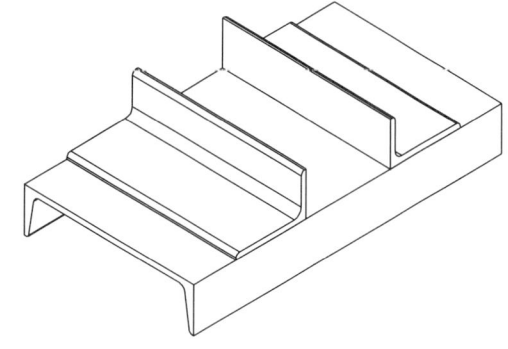

BOMKEY ITEM QTY NAME MATERIAL VENDOR NOTE
4282 1 1 U-SHAPE - AISC - C 12 X 25_1 ASTM A36
4283 2 2 ANGLE STEEL - AISC - L 5 X 3 X 3_8_1 ASTM A36

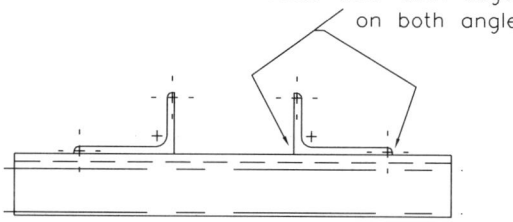

Fillet weld both edges on both angles

14

MECHANICAL DESKTOP POWER PACK 3D CONTENT: PREMODELED PARTS AND HARDWARE COMPONENTS

Many basic engineering parts, components, and hardware items such as fasteners, shafts, bushings, structural shapes, and bearings, are available in Mechanical Desktop. Once these parts are defined and placed into the drawing, they can be inserted into an assembly just like any other created parts. Let's walk through an example of applying these predefined components, starting with fasteners.

Step 1 Type **AMSCREW3D** ↵
or click the icon.

The Select a Screw dialog box appears.

218 | CHAPTER 14

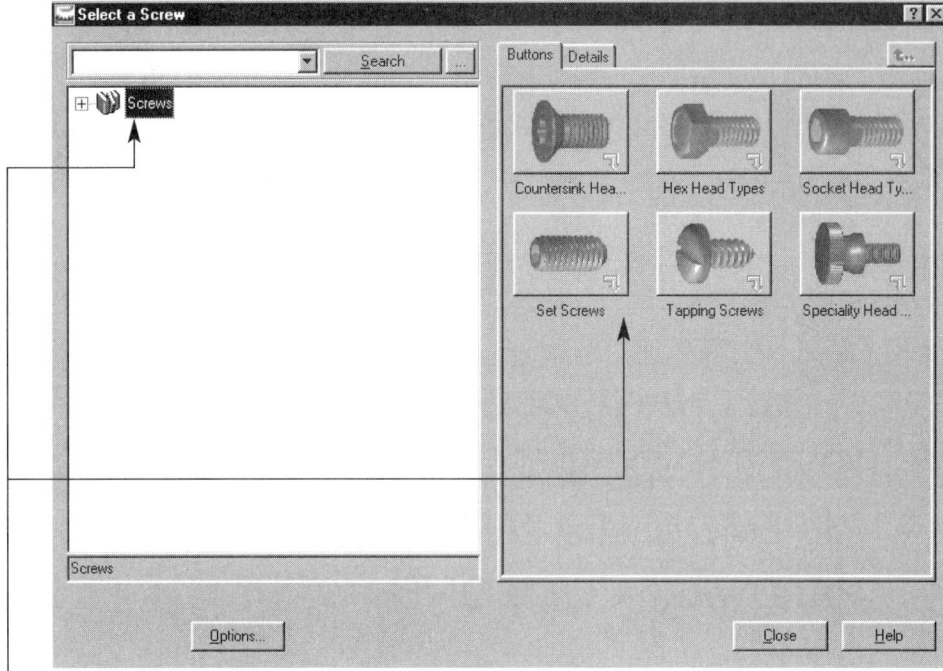

Click on one of the **screw styles** in the list or on one of the graphic representations.

Step 2 Further refine your selection by clicking on a **screw type** in the list or one of the graphic representations.

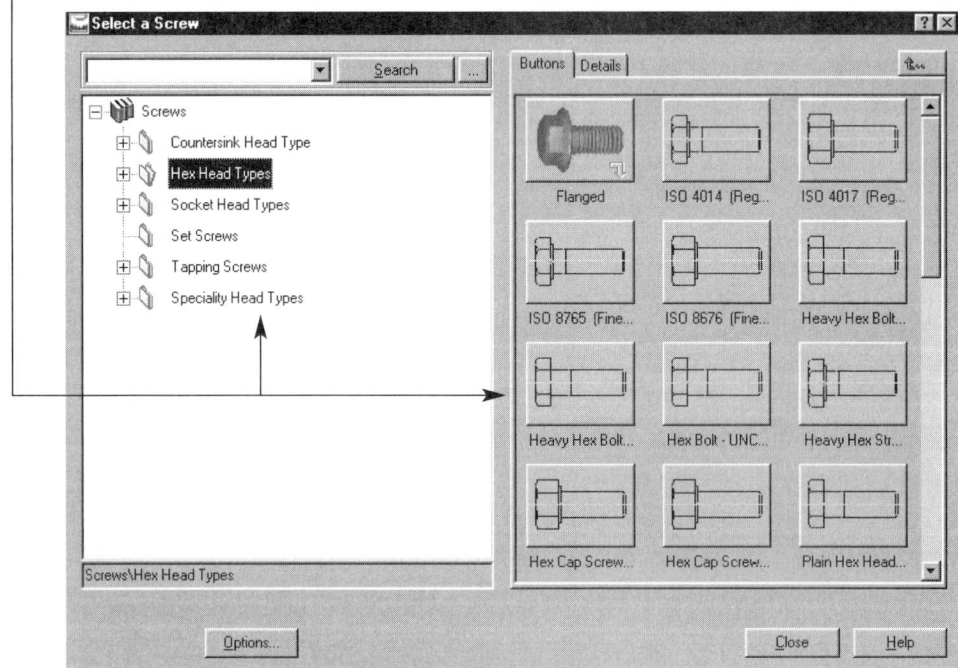

Step 3 Select first point [Concentric/cYlinder/two Edges]:
(Click an initial point or property on the screen.)
Select second point [Concentric/cYlinder/two Edges]:
(Click a second point on the screen.)

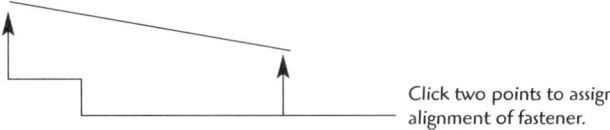

Click two points to assign alignment of fastener.

Step 4 The screw **size** box appears.

Click the size of the screw fastener.
Click **Finish**.

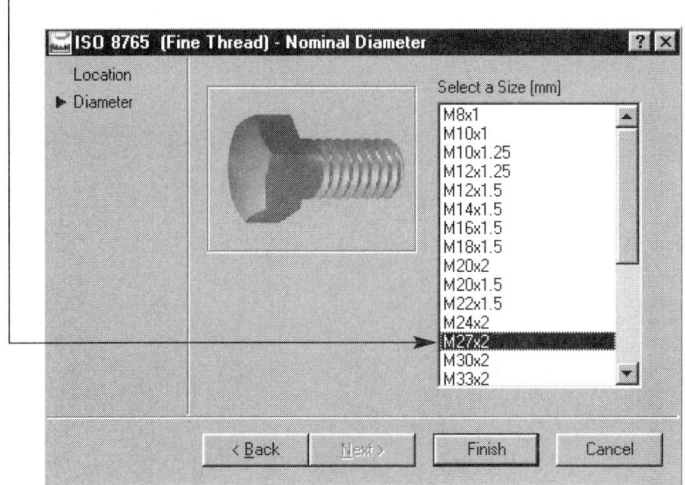

Step 5 Drag or type the length of the screw fastener on the model screen.

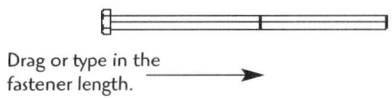

Drag or type in the fastener length.

If you type in a length, press ↵.

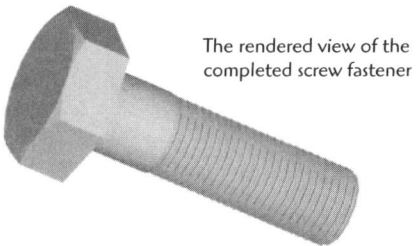

The rendered view of the completed screw fastener

You can change the representation of the fastener by typing **AMSTDPREP** ↵.

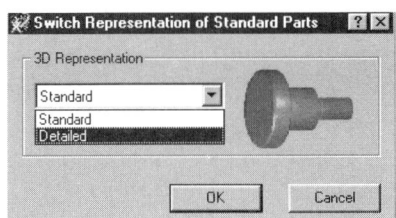

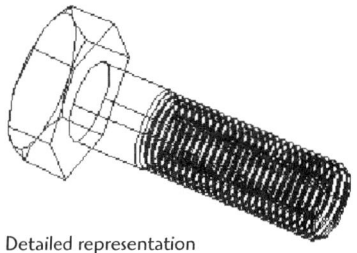

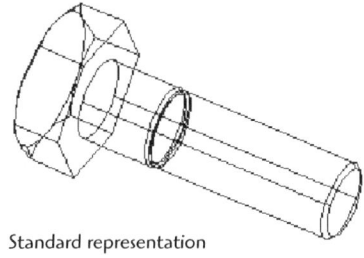

Detailed representation

(The detailed representation will use more of your computer memory.)

Standard representation

Now let's look at structural shapes.

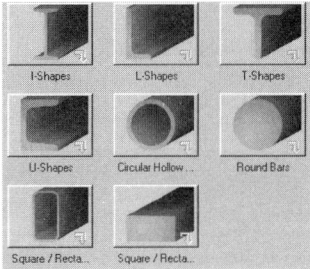

Step 1 Type **AMSTLSHAP3D** ↵
or click the icon.

The Select a Steel Shape dialog box appears.

Step 2 The following steel shapes are available:

- Angle
- W, HP, M and I beam
- U-channel iron
- Square and rectangular
- T and Z iron
- Pipe and solid round

Select a steel shape from the display or the list.

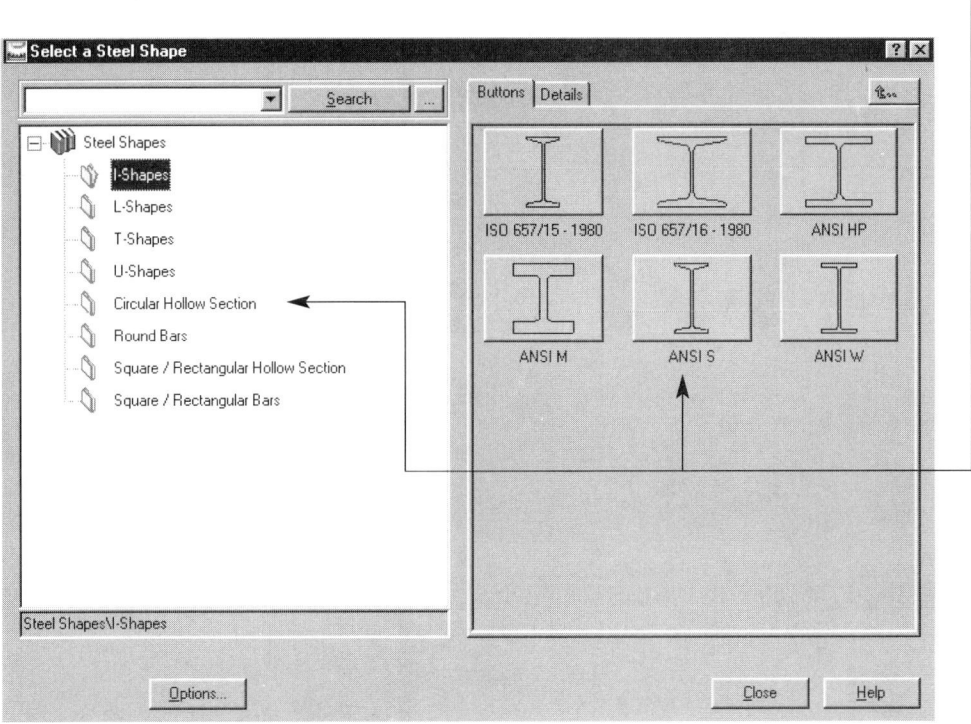

Step 3   `Select first point [Concentric/cYlinder/two Edges]:`
            `Select second point [Concentric/cYlinder/two Edges]:`

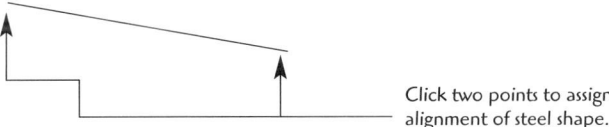

Click two points to assign alignment of steel shape.

Step 4 Refine your selection.

In the case of W beams, the depth × weight per/ft is the required selection.

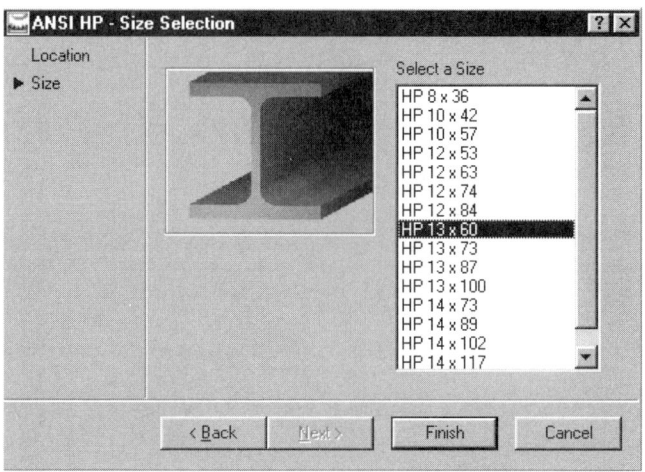

Step 5 `Drag size [Dialog/Associate to/Equation assistant]:`
(Drag or type in the length of the structural shape.)

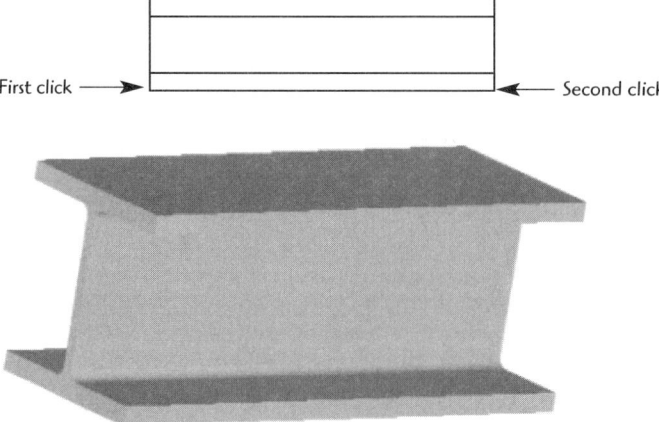

The completed structural shape ready to be added to the assembly catalog

MECHANICAL DESKTOP POWER PACK 3D CONTENT | 223

For a Deeper Understanding about
Premodeled Parts and Hardware Components

The development process is much the same for the many other shapes that are available. Let's look at some of the other options.

Fastener Options

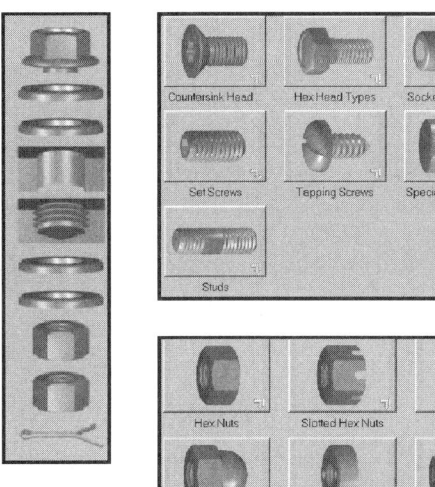

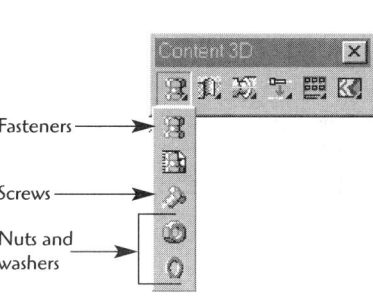

Fasteners

Screws

Nuts and washers

Shaft Components

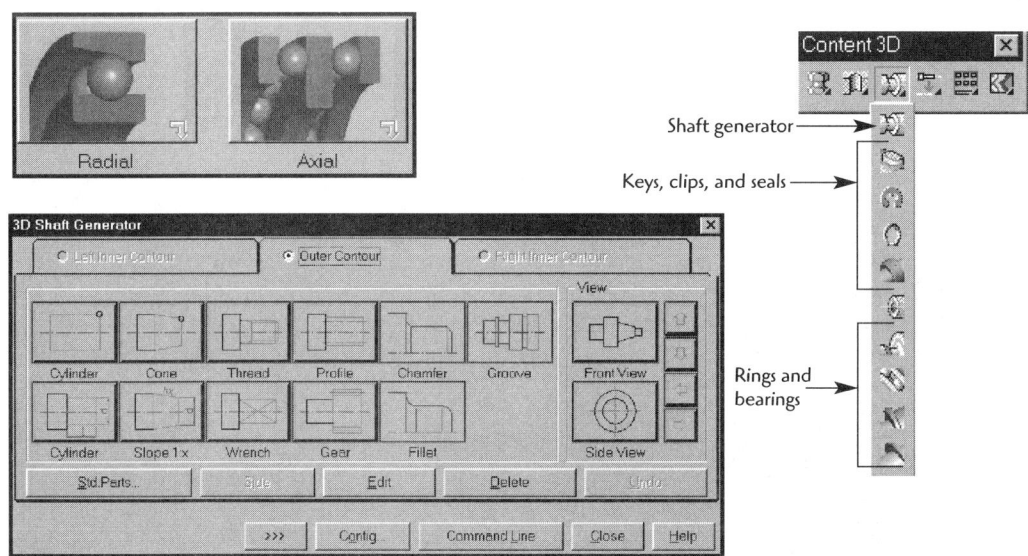

Shaft generator

Keys, clips, and seals

Rings and bearings

CHAPTER 14

Structural Components, Pins, Rivets, Plugs, and Bushings

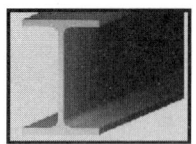 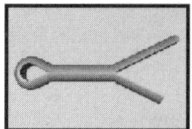 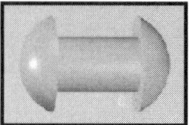 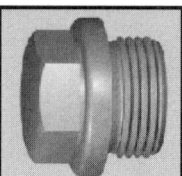 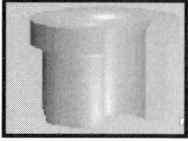

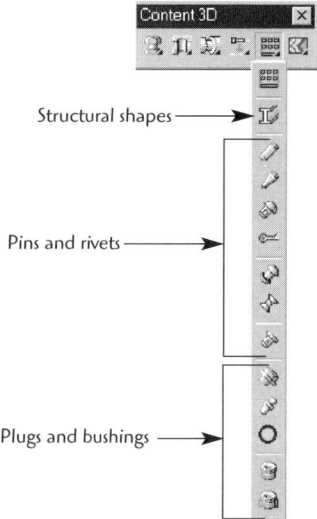

Structural shapes

Pins and rivets

Plugs and bushings

Standardized Holes

Mechanical Desktop has standardized holes for ISO, ANSI, DIN, and user-defined holes.

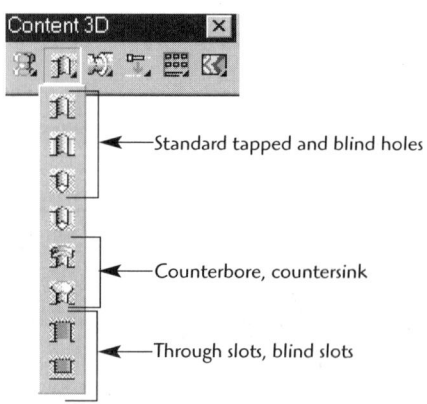

Standard tapped and blind holes

Counterbore, countersink

Through slots, blind slots

MECHANICAL DESKTOP POWER PACK 3D CONTENT | *225*

The development of these holes is similar to that of using the AMHOLE command; however, the dimensions of the hole will be held to standard sizes. Let's walk through an example of a counterbore hole.

Type **AMCOUNTB3D** ↵

or click the icon.

The counterbore hole standard box appears.
Select a counterbore type.

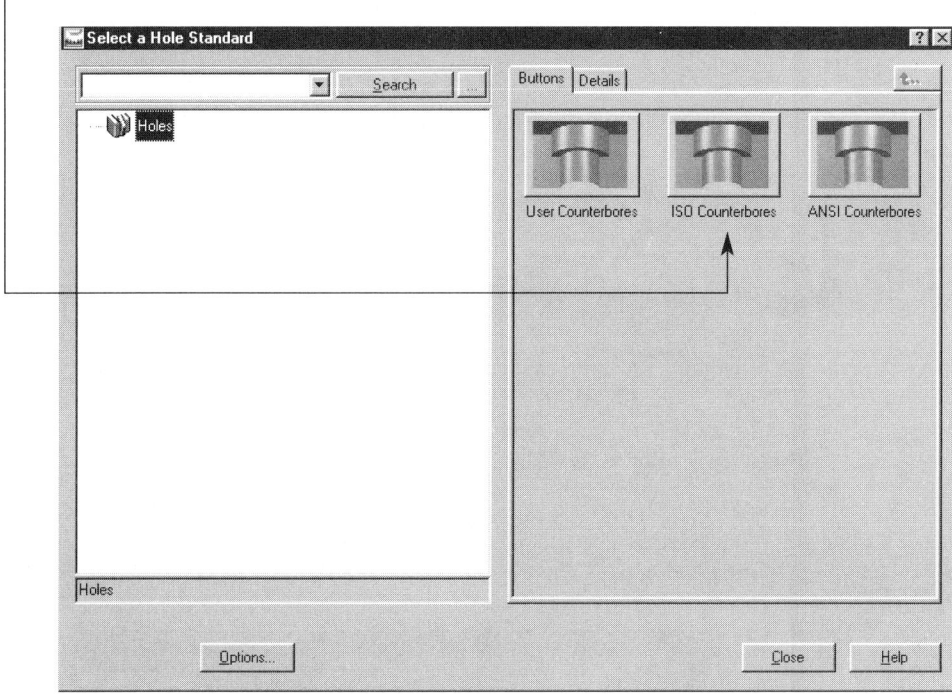

The Select a Screw box appears.
Select a specific screw style.

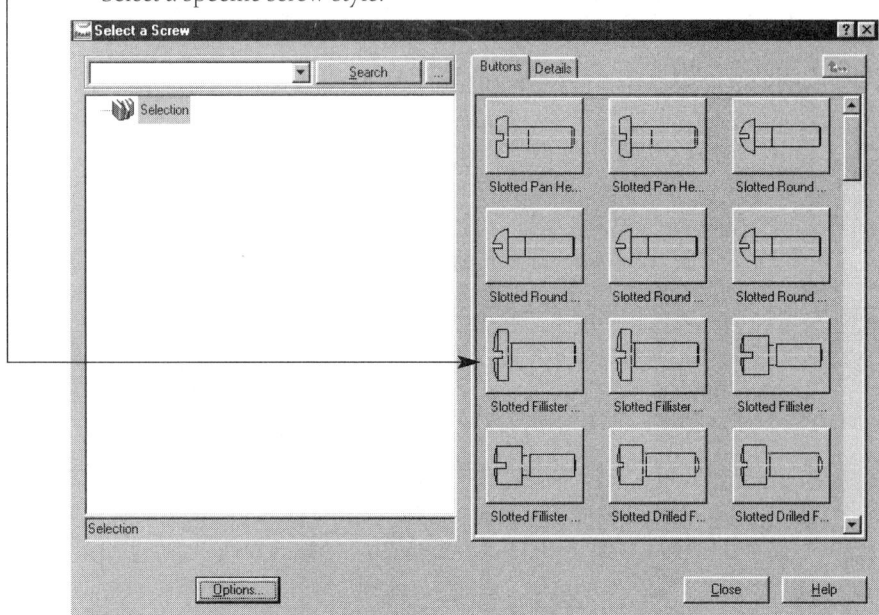

The Screw Connection - 3D dialog box appears.

This handy box allows you to select and specify all components of the connection. In this case you are set to specify the Slotted Fillister Head Cap Screw.

Select the component size.

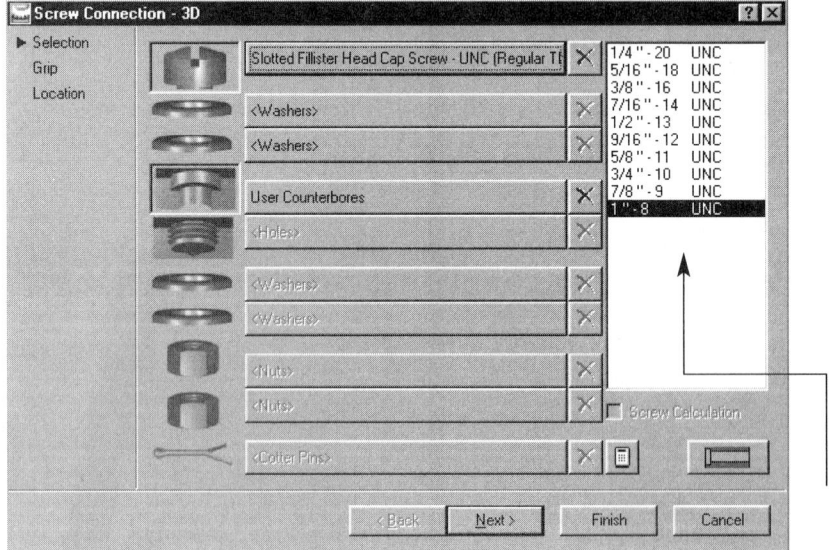

Next, the placement box appears.
Select a hole position method.

- Select first edge or planar face:
- Select second edge or planar face
- Enter hole location
- Type in first edge distance and press Enter to confirm.
- Type in second edge distance and press Enter to confirm.
- Hole in termination: <thru> press Enter

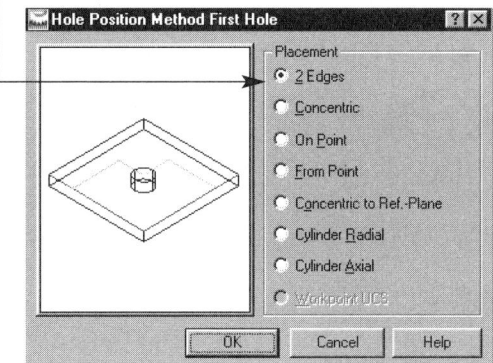

The Screw Assembly Location - 3D dialog box is used to locate the hole assembly components.

In the case of a single hole, simply click **Finish**.

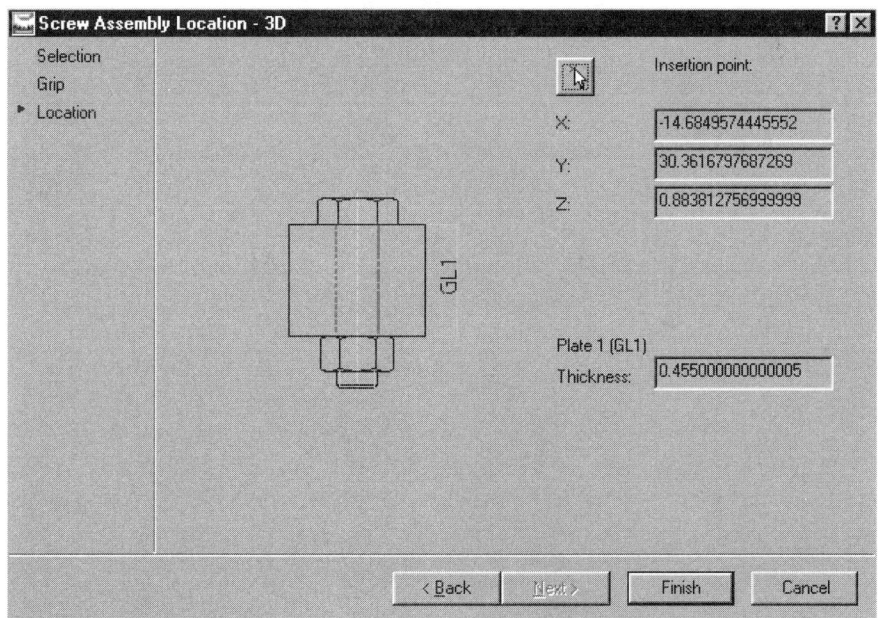

The Enter Values dialog box appears.

The standard dimensional sizes are listed for the hole diameter, counterbore diameter, and the counterbore depth. If needed the values in the *Value* column can be adjusted.

Click **OK**.

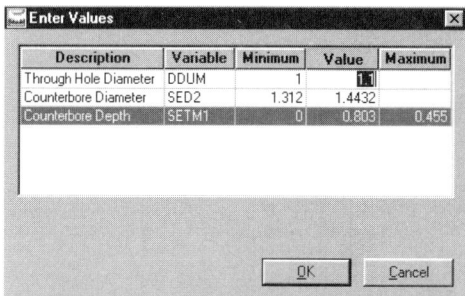

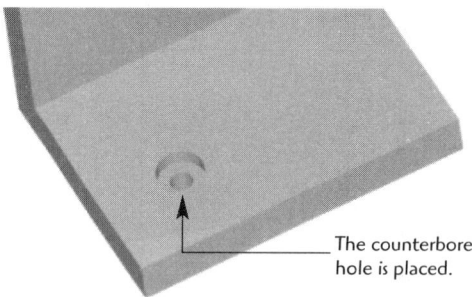

The counterbore hole is placed.

Drawing Project 14-1

Using Power Pack Components

For this project you will again produce an assembly drawing similar to the following drawing, including the bill of materials, but you can now use the power pack predefined parts for the hex bolts. Create an exploded view scene and bill of materials for the assembly. Place the exploded view of the assembly along with orthographic views of each part into the final drawing mode view. The drawing should be detailed enough that a reader could produce the part. When you have finished, print out a copy of the drawing for your instructor.

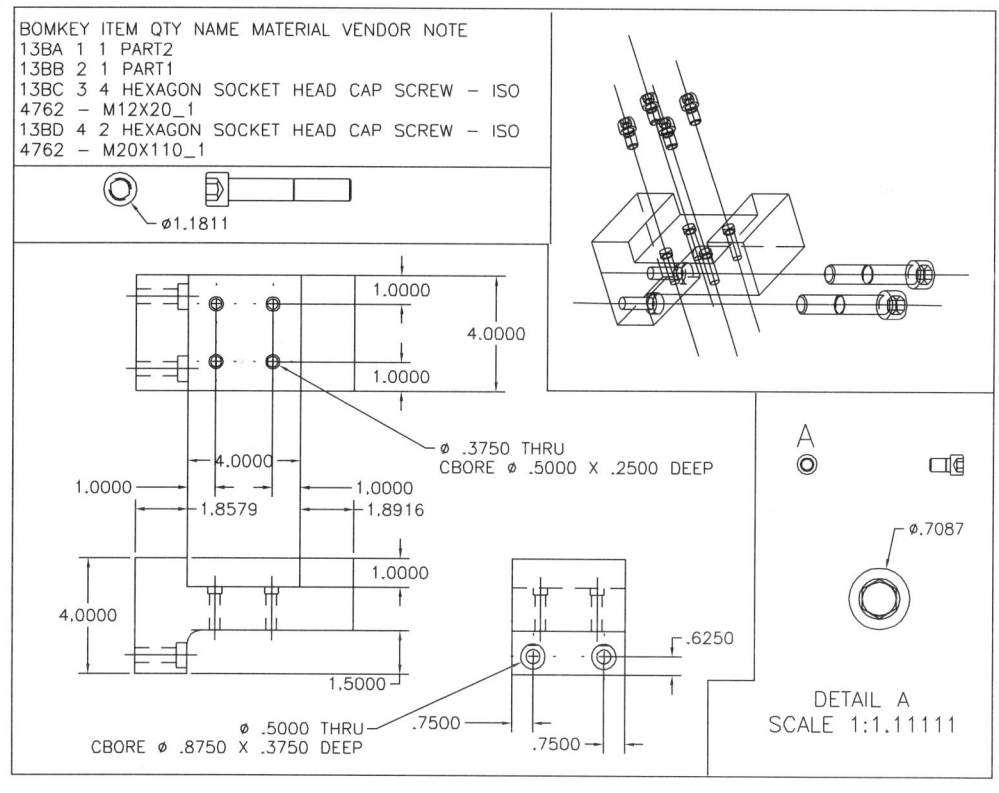

Drawing Project 14–2

Using Power Pack Structural Components

For this project you will again produce an assembly drawing similar to the following drawing, including the bill of materials, but you can now use the power pack predefined parts for the structural components. Create an exploded view scene and bill of materials for the assembly. Place the exploded view of the assembly along with orthographic views of each part into the final drawing mode view. The drawing should be detailed enough that a reader could produce the part. When you have finished, print out a copy of the drawing for your instructor.

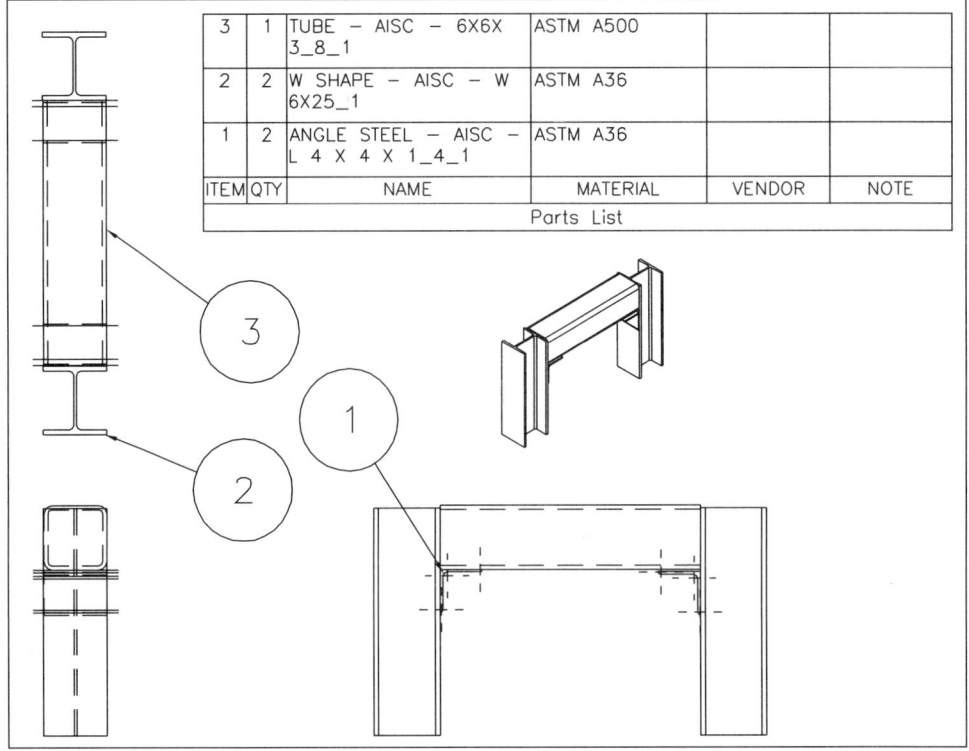

15

FINITE ELEMENT ANALYSIS

When parts are designed and put into service we certainly hope they have been tested and verified to have the integrity for their service life. In other words, we hope they will not fail or break while in service. Testing and analysis helps designers and engineers verify that a part has sufficient mechanical properties. As a part design is developed, improvements are incorporated in the design. Finite element analysis (FEA) aids in this improved design development. Although not a final or conclusive test of part integrity, high-stress areas are calculated, analyzed, and displayed for the designer. The designer can then use this information to improve the design.

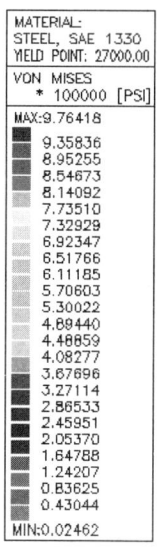
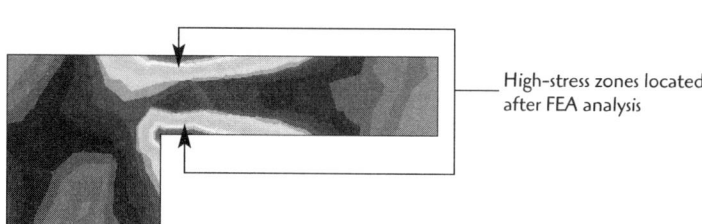

High-stress zones located after FEA analysis

The first step of the FEA process requires the application of restraints and forces to the modeled part. Next, a mesh is applied to the part so that the stress analysis can be performed on each meshed cell. Finally, the solution is displayed with a variety of display methods.

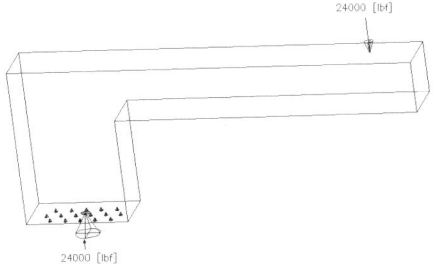

First step: Application of restraints and forces

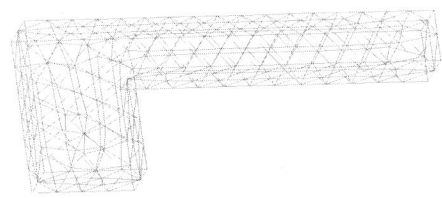

Second step: Application of the mesh

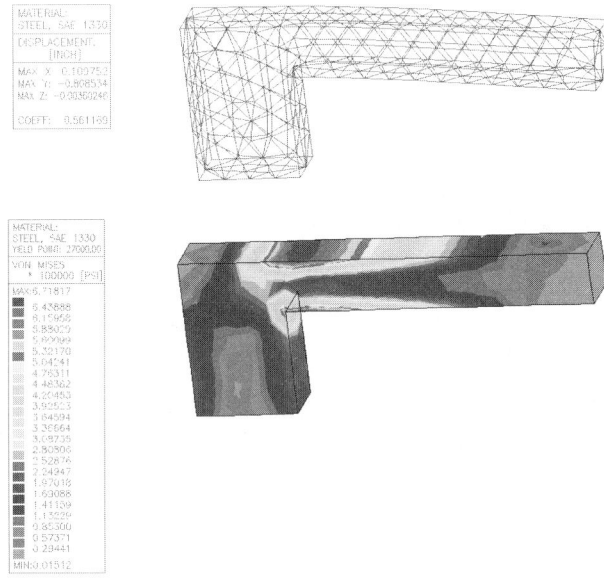

Third step: Display of results
The top model shows displacement.
The bottom model displays cell stress.

Part Analysis (FEA) with Mechanical Desktop

Step 1   Type **AMFEA3D** ↵.

```
Select 3D-body:
```

Select the part.

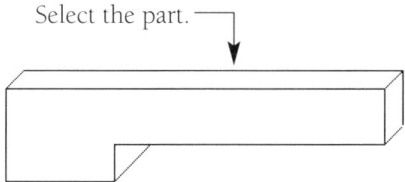

You first need to anchor a portion of the part. Click the anchor button.

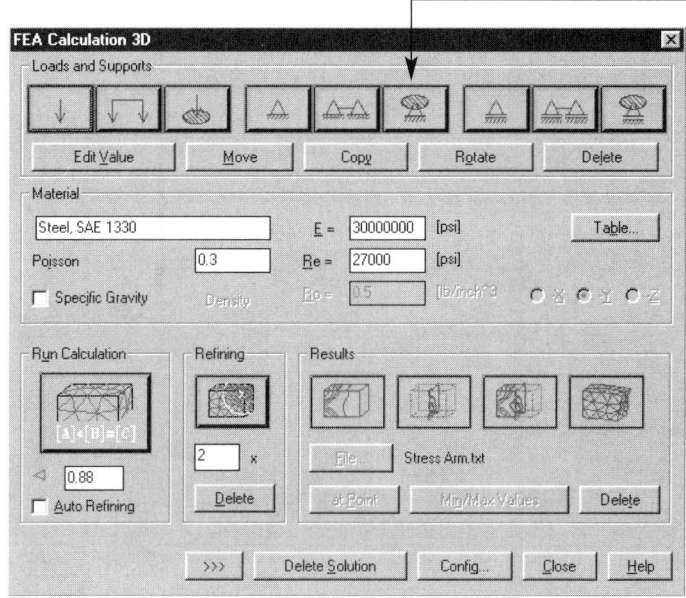

Step 2   Pick on an edge near the face to be anchored, then ↵.

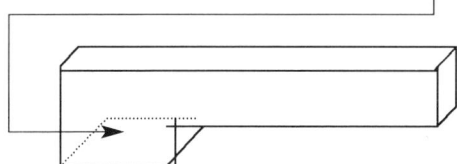

Step 3 Select **Whole Face**.
Drag the select lines to the center of the face and select.

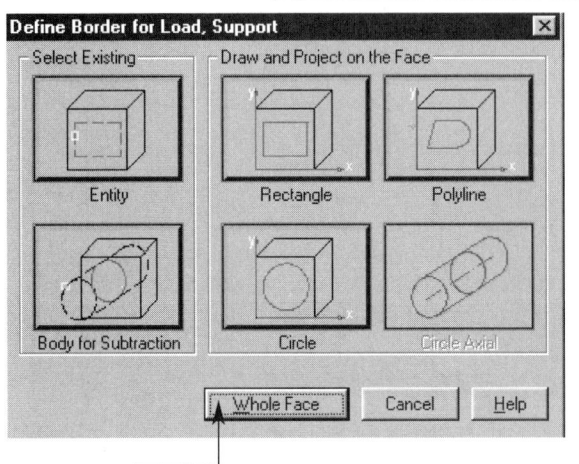
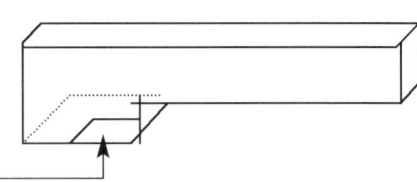

Step 4 Pick the load button.
Select the load face, then ↵.

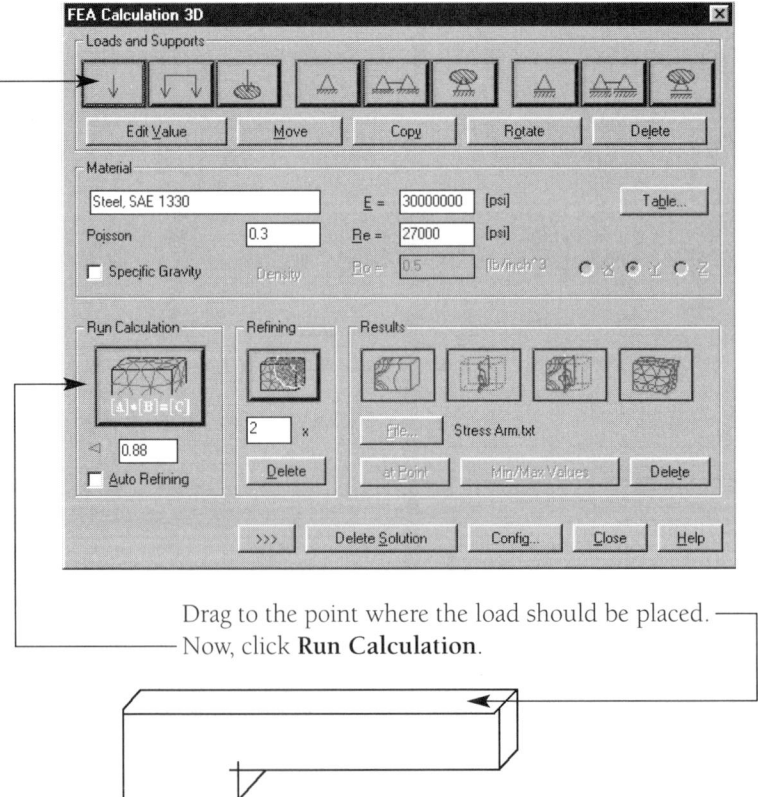

Drag to the point where the load should be placed.
Now, click **Run Calculation**.

FINITE ELEMENT ANALYSIS | **235**

Step 5   The part will have a *mesh* applied to it so that *mesh cells* can be created to calculate and display stress regions. The next step is to move the solution off the original part model.

> `Specify a base point or displacement:` *(Click on the part and drag it below the original part.)*

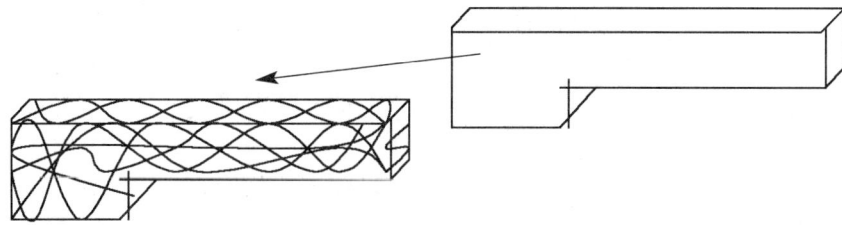

Step 6   After selecting Run Calculation, you wait for a solution to be generated. Follow one of these steps to display the solution graphically.

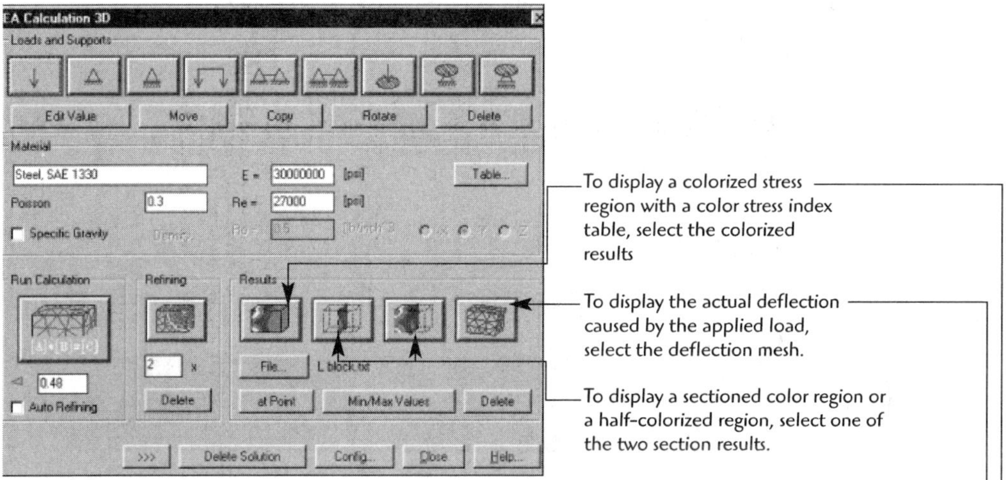

Display options

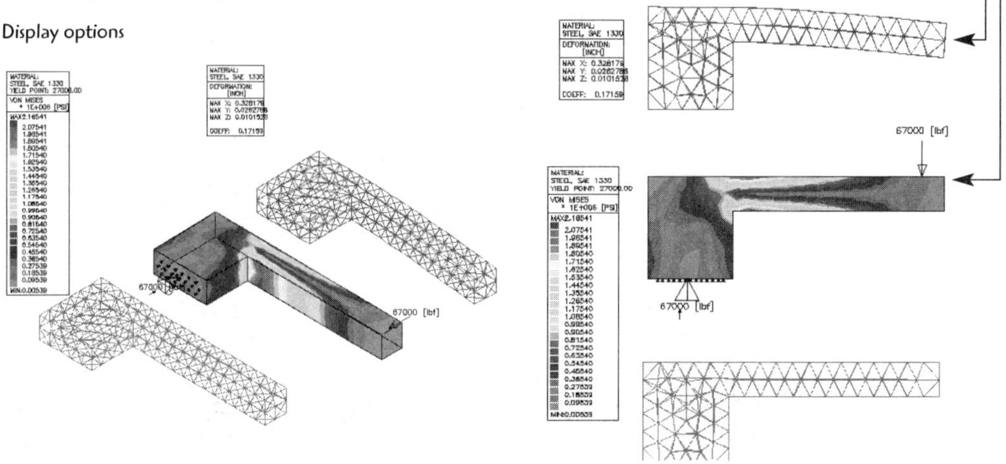

For a Deeper Understanding about Part Analysis

For complete accuracy of your analysis, you need to set the correct material for the analysis.

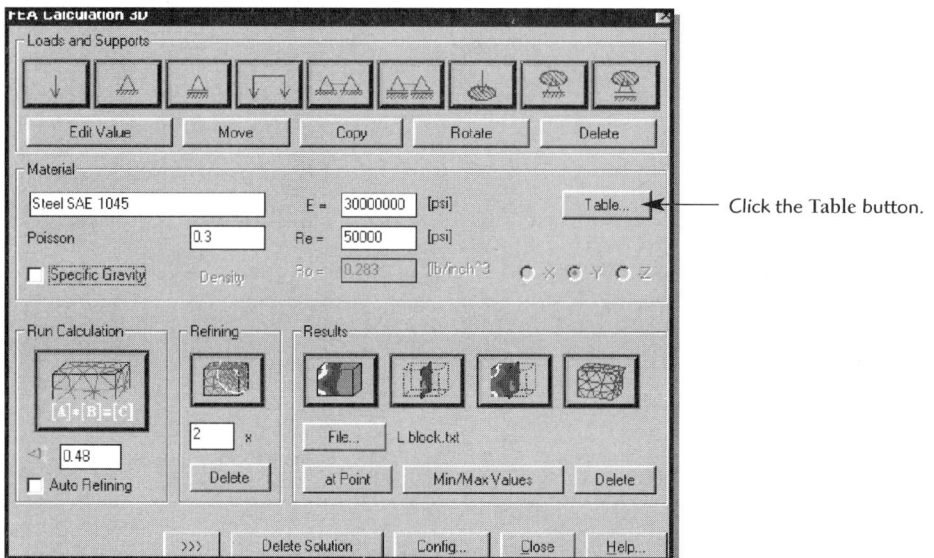

Click the Table button.

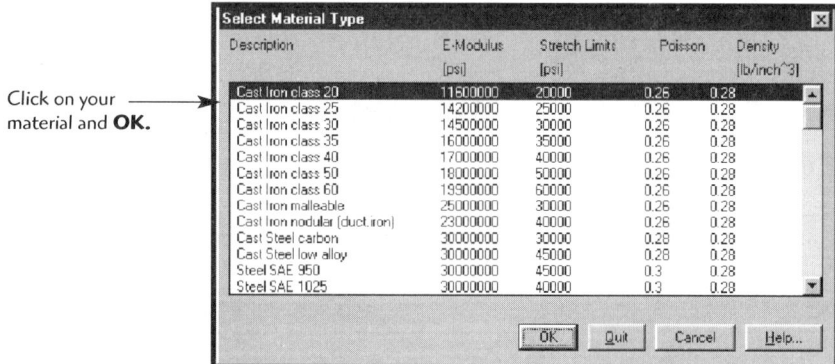

Click on your material and **OK.**

The material properties will change to match the material you have selected. You should now re-solve the model by clicking Run Calculation again.

FINITE ELEMENT ANALYSIS | 237

You can use the **Refining** option if you have more computing power on your computer. The mesh generated will be smaller and finer, resulting in a more accurate FEA model.

It is also helpful for the analysis to refine the application of the loads and supports.

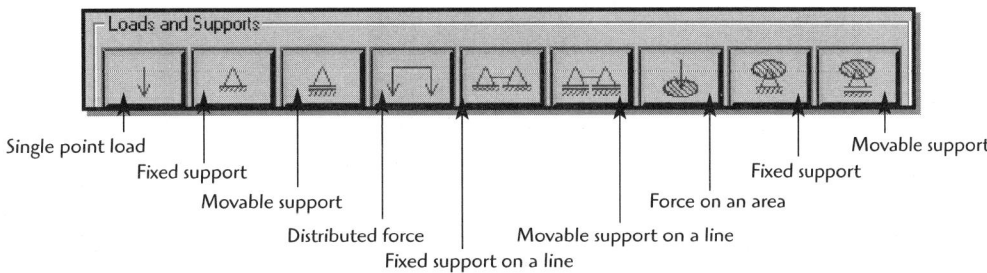

Single point load
Fixed support
Movable support
Distributed force
Fixed support on a line
Movable support on a line
Force on an area
Fixed support
Movable support

OBTAINING THE SOLID MASS PROPERTIES FOR PARTS

Once a model is completed, Mechanical Desktop allows you to extract the solid mass properties for that part. You can assign a material (material density) and save the data into a file. This file can be attached to the drawing or printed out.

Step 1 Model and fully constrain the part.

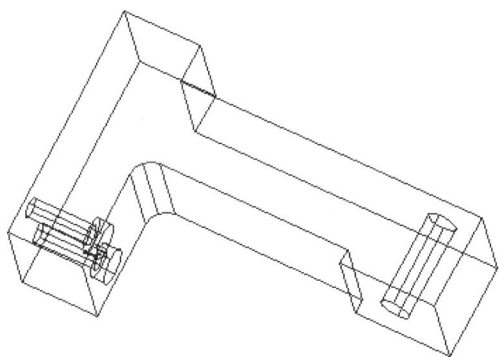

Step 2 Type **AMMASSPROP** ↵
or click the icon.

The Assembly Mass Properties dialog box appears.

You must select the material or enter the material density. All materials have a unique density. An exact density for materials can be found in many physics, metallurgy, and engineering manuals. Here are some common densities (in cubic inches) for your use:

- Low-carbon steel = .2839
- Medium carbon steel = .283
- Stainless (grade 304) = .290
- Copper = .324
- Aluminum = .098
- Magnesium = .066
- Nickel steel = .322
- Titanium = .1628

(*Note:* The dialog box gives some densities in g/cm^3.)

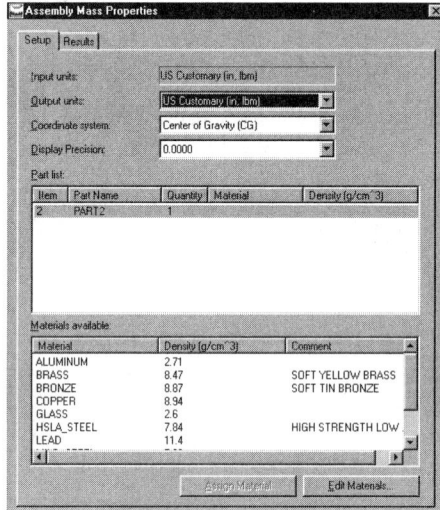

This program is set up to use US standard, SI (metric), ISO, and ANSI units.

Once you select a material, click on **Edit Materials.** Adjust the density value if necessary, then click **OK.** Now, click the **Results** tab, then click **Calculate.** Select **Export Results** to save the results into a file. (*Note:* The Save box with the .mpr extension appears. Save this file to a location that you will remember.)

Click **Done**.

Step 3 Type **MTEXT** ↵.

Window an area on the drawing mode screen.

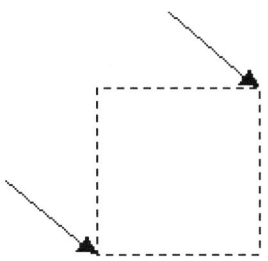

Click **Import Text**.

Type `*.mpr` in the file name area and ↵. (The mass property file that you saved should be visible. Select the file.)

Click **OK**.

Position the mass properties file on the drawing. You can use the SCALE command to adjust the size of the text to better fit on the drawing screen.

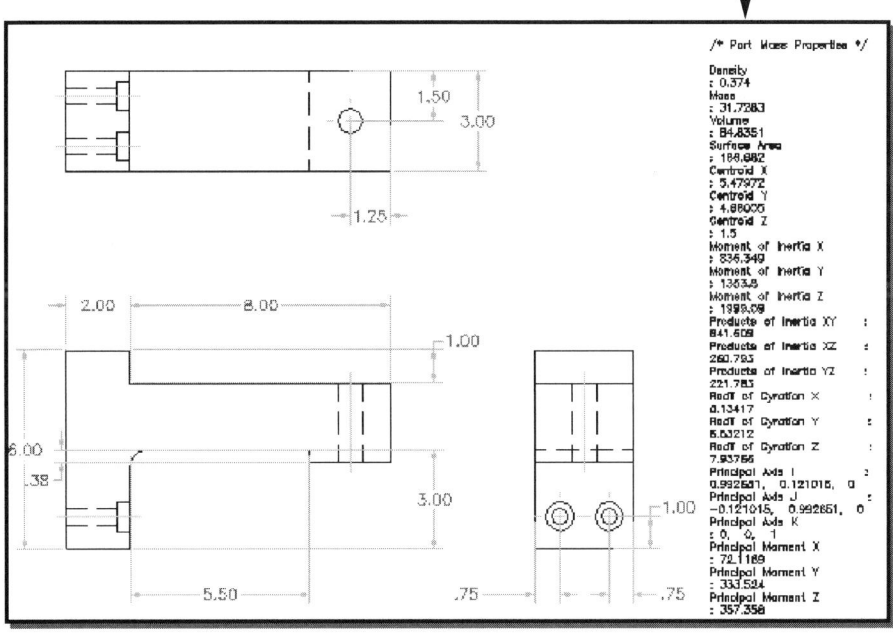

Drawing Project 15–1

Mass Properties Calculation

Model the following part below to the exact dimensions indicated. Use Revolve and add the fillets as needed to complete the model. The part is made from aluminum. Save the mass properties file and attach it to your final drawing. Print out a copy of the drawing for your instructor.

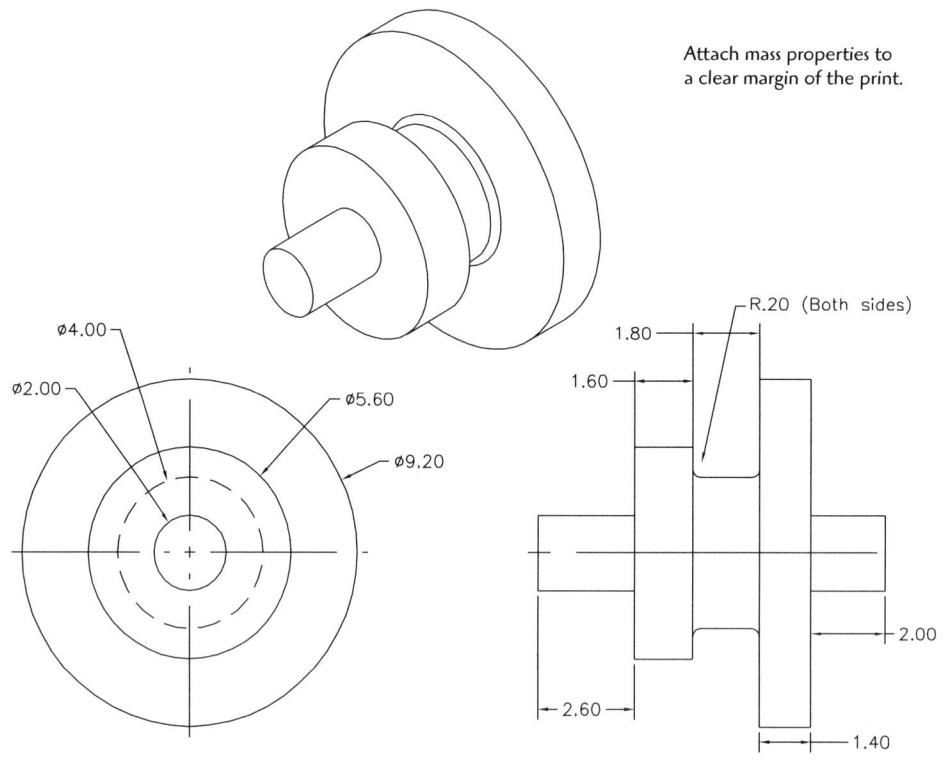

Attach mass properties to a clear margin of the print.

Drawing Project 15-2

Mass Properties Calculation

Model the following part below to the exact dimensions indicated. Use Extrude, and add the fillet and holes as needed to complete the model. The part is made from low-carbon steel. Save the mass properties file and attach it to your final drawing. Print out a copy of the drawing for your instructor.

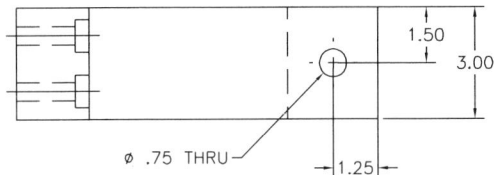

Attach mass properties to a clear margin of the print.

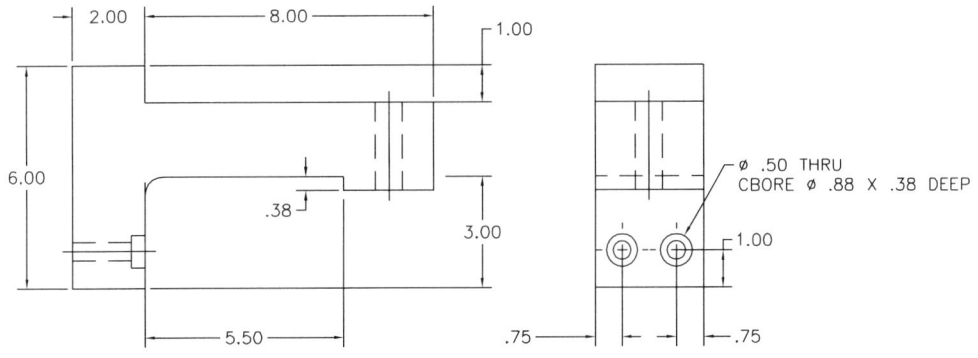

Drawing Project 15-3

Vice Assembly

For this special challenge project you will need to use your creativity. You will produce an assembly drawing similar to the following drawing from other parts you have developed, but there are no dimensions. You must produce and refine a working design and develop your own dimensional sizes. You must also test your design for interferences. Create an exploded view, scenes, and bill of materials for your final design. Place the exploded view of the assembly along with orthographic views of each part into the final drawing mode view. The assembly drawing should be detailed enough that a reader could produce the part. When you have finished, print out a copy of the drawing for your instructor.

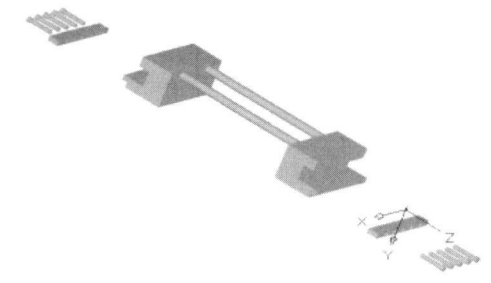

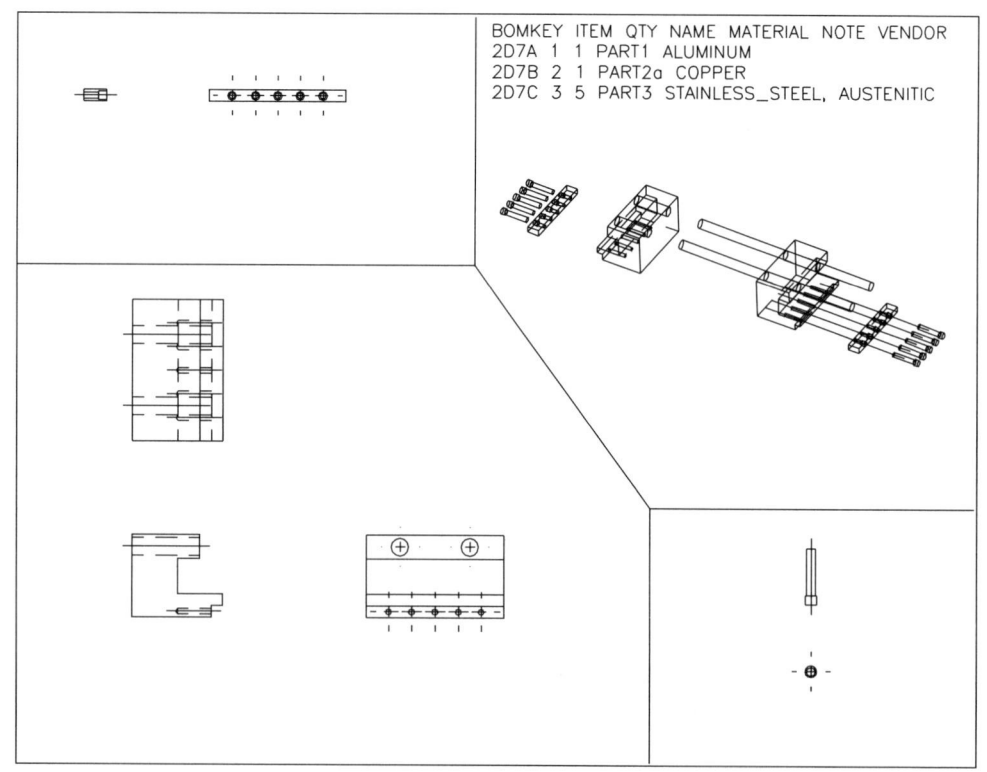

BOMKEY	ITEM	QTY	NAME	MATERIAL	NOTE	VENDOR
2D7A	1	1	PART1	ALUMINUM		
2D7B	2	1	PART2a	COPPER		
2D7C	3	5	PART3	STAINLESS_STEEL, AUSTENITIC		

USING THE MECHANICAL DESKTOP INTERNET TOOLS

With the power and flexibility of the Internet you can:

- Create web pages that display detailed drawings and models.
- Communicate with project engineers on drawing updates.
- Collaborate with other designers.

Note: The following assembly is a **web page browser view**, not a Mechanical Desktop view. The view looks like a typical Mechanical Desktop view but has been saved as a **.DWF** (drawing web format) file.

Note the following right-click options available in a .DWF file:

- Pan
- Zoom
- Zoom Rectangle
- Layers
- Highlight links (URLs)
- Print drawing

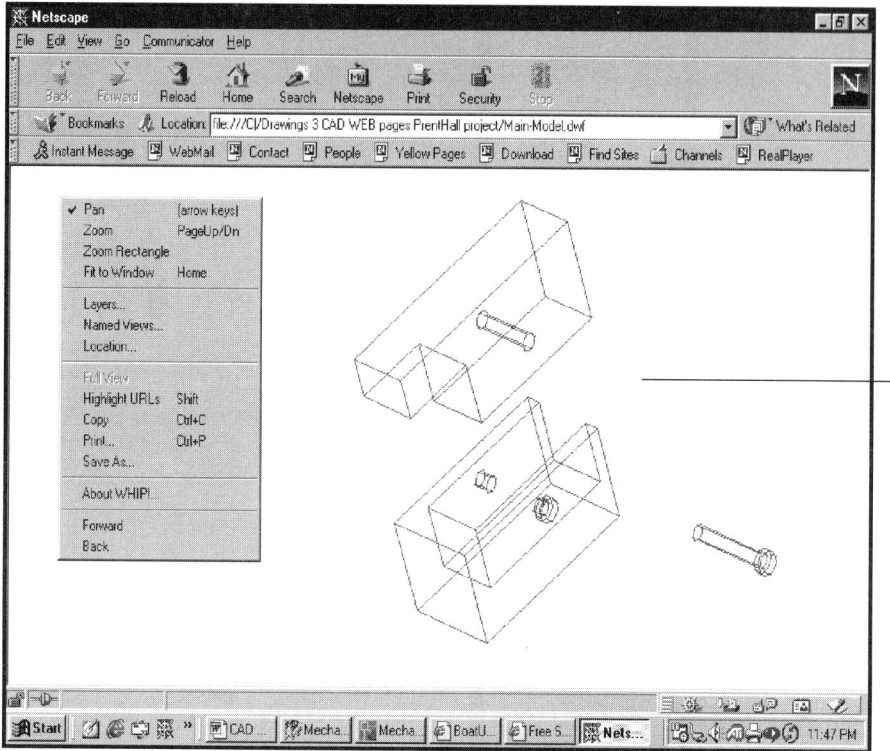

Clicking on one of the hyperlinks will link you to a detail view page.

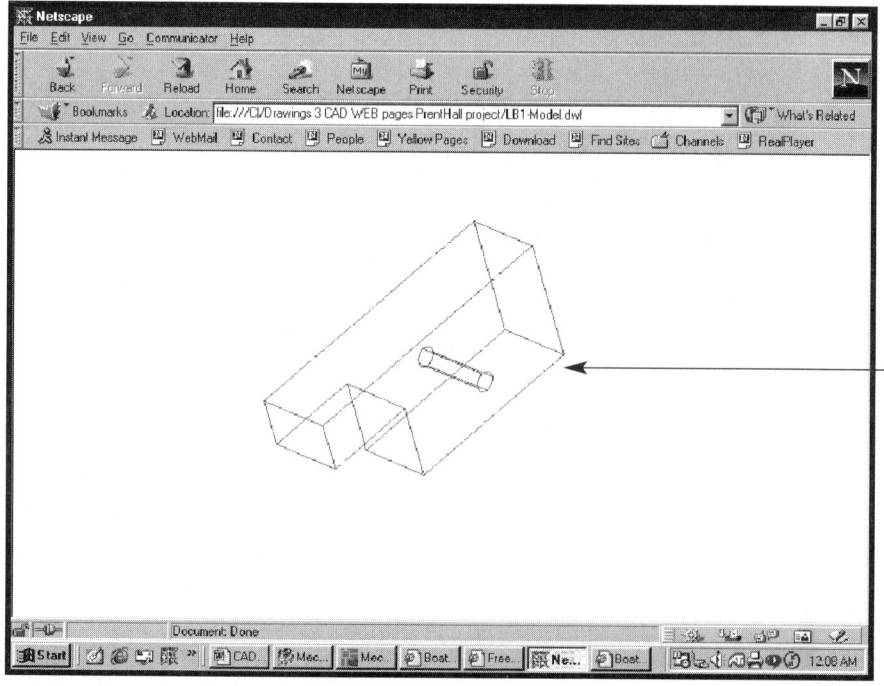

The advantage of creating CAD web pages is to provide a drawing design resource that can be accessed from around the world. Viewers of your page can examine, zoom, rotate, and hyperlink on details of your drawings and models. The following diagrams outline the basic process of creating CAD web pages.

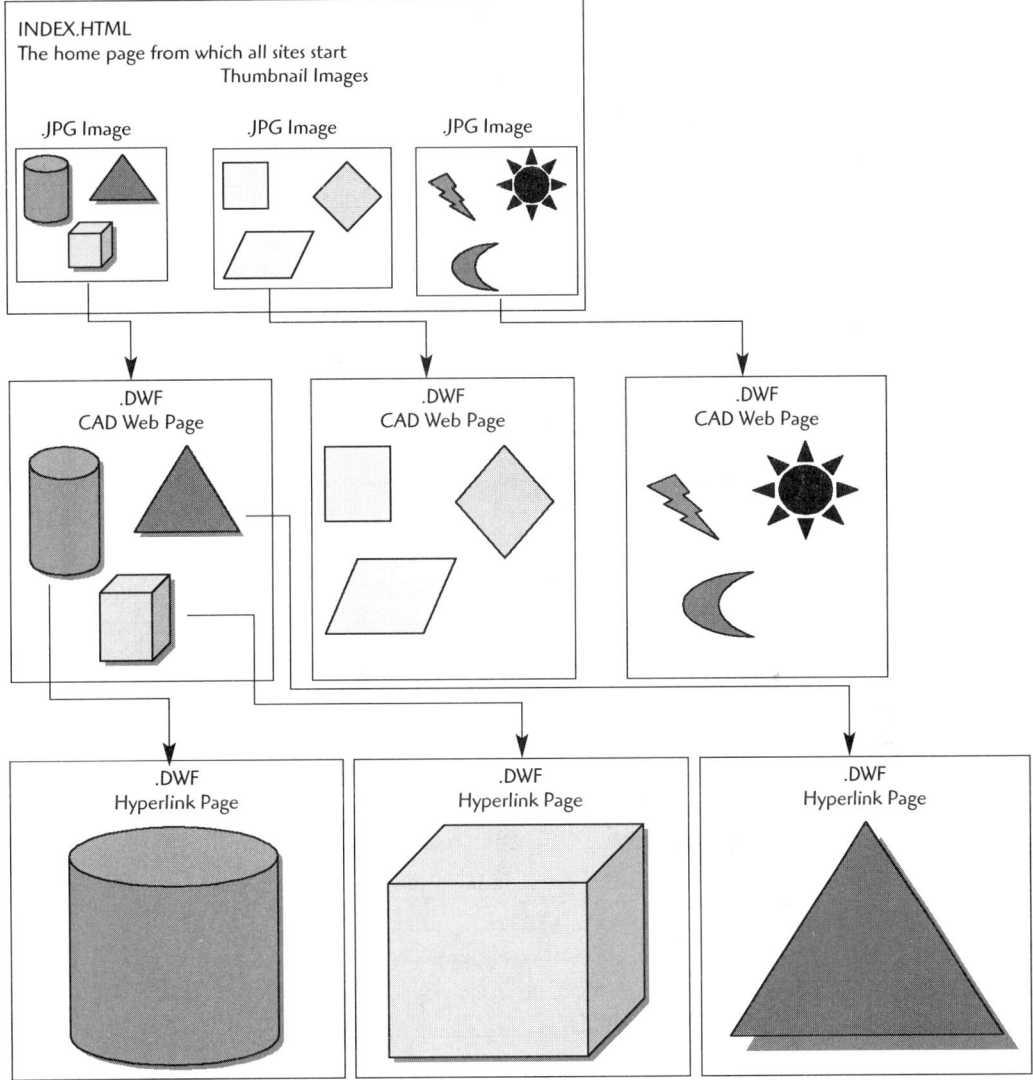

246 | CHAPTER 16

Let's step through the first (manual) method for creating CAD web pages

Step 1 Create the index page.

> **INDEX.HTML**
> The INDEX.html page is created with a web page editor such as *Netscape Composer, Front Page,* or *Dreamweaver*

Step 2 Create the CAD pages.

- Create or open the main drawing in AutoCAD.
- Use **WBLOCK** to create a detail of each part.
  1. Use **Select objects** to select the detail.
  2. Use a unique file name for each detail.
  3. Save in your CAD web page file.

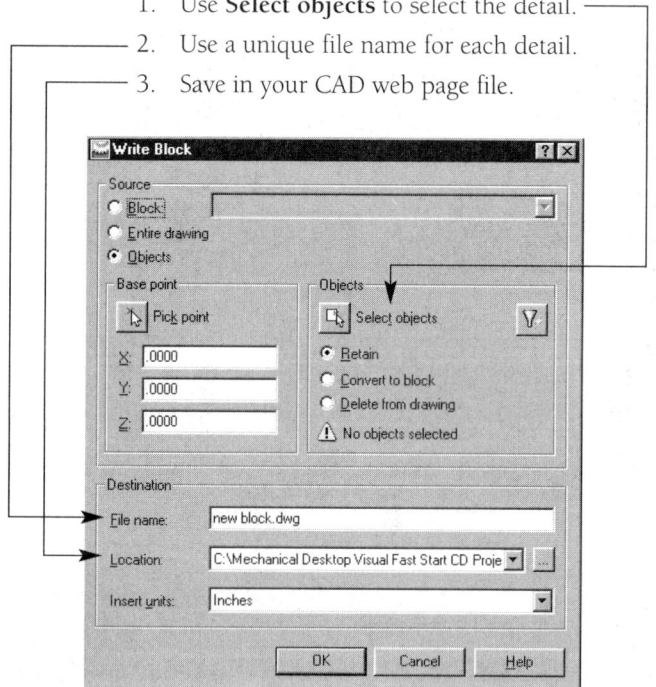

- Open each of the WBLOCKED parts and use **PLOT**, then select **DWF ePLOT.pc3** (to export as a .DWF file).
- Close and return to the main drawing.

USING THE MECHANICAL DESKTOP INTERNET TOOLS | 247

Step 3 Create the hyperlinks

- Type **HYPERLINK**↵, then select one of the details ↵.
- Browse to the .DWF detail and click **OK**.
- Once all the details are hyperlinked, use PLOT, then select DWF ePLOT.pc3 to export the main drawing as a .DWF file.

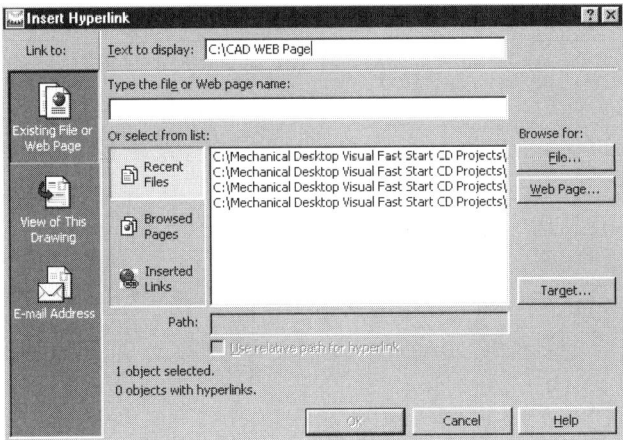

The **second method** for producing a CAD web page is to use the **Publish to Web** option in the File pull-down menu. This method provides a more automated system of producing a CAD web page.

- Click **File**, then **Publish to Web**.
- Click **Create New Web Page**.
- Click **Next**.

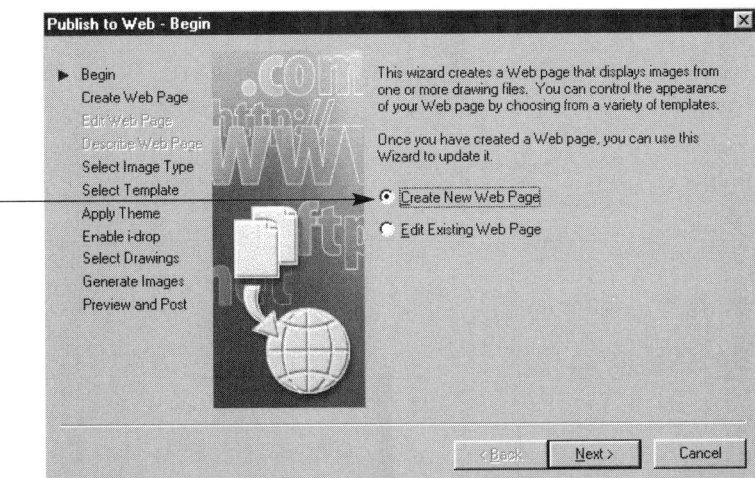

- Type in a heading name for your page and a page description.
- Click **Next**.

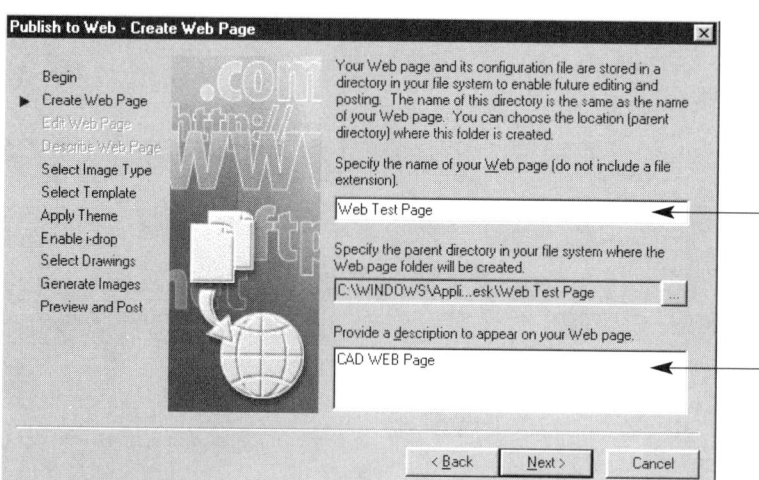

- Select thumbnail images layout.
- Click **Next**.

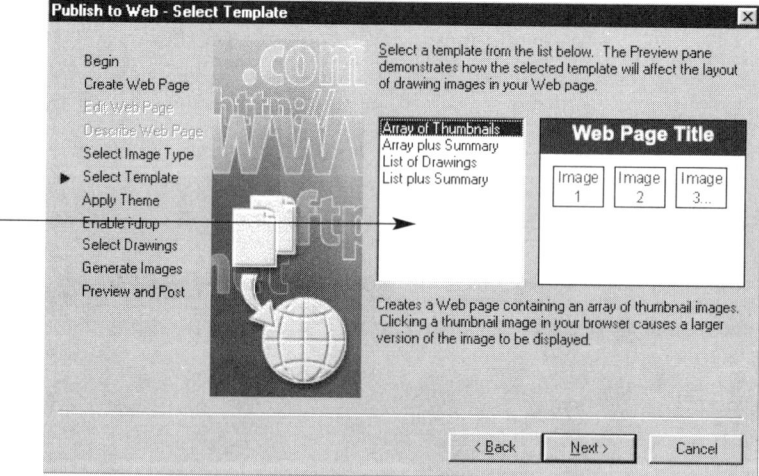

- Select **JEPG**.
- Click **Next**.

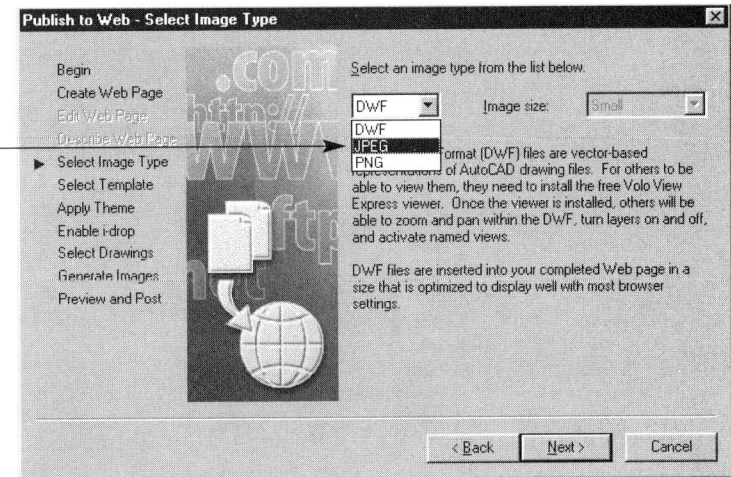

- Select a theme and click Next, and Next again.

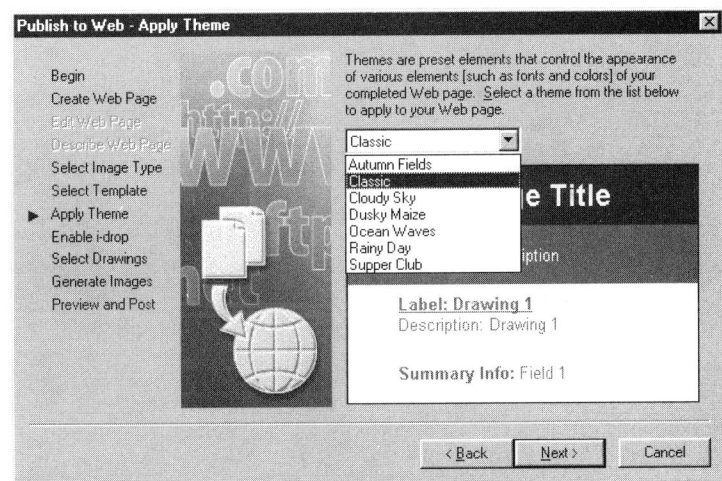

This important step requires you to locate and add each drawing you want on the home page.

- Use the ... button to navigate to drawing files, then click **Add**.
- Click **Next**.

- Click the **Regenerate images** button, then **Next**.

    (Be patient. This process will take a while as it creates each .JPG and .DWF file.)

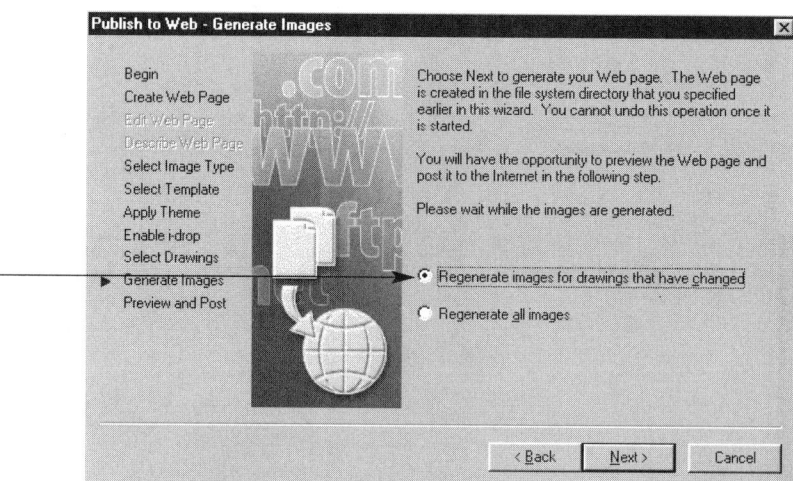

- Click the **Preview** button to see the page. (You would use **Post Now** if you had a site location defined.)

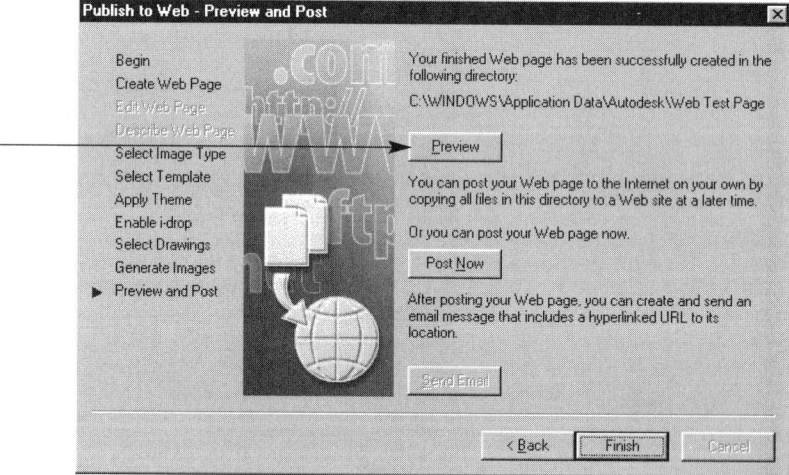

Here is the web page, complete with links to the drawing web format pages.

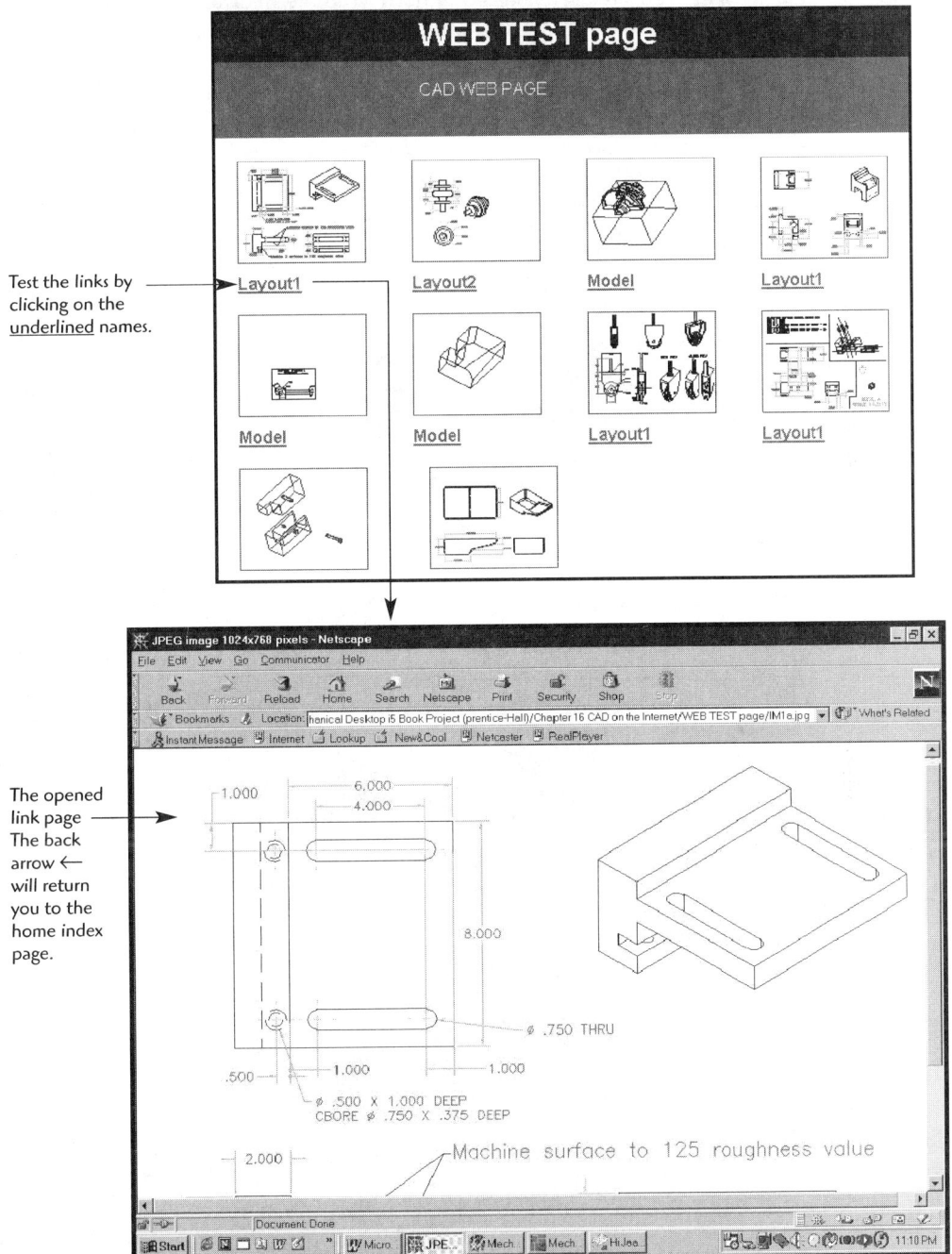

Test the links by clicking on the underlined names.

The opened link page. The back arrow ← will return you to the home index page.

Click the **Finish** button as the final step in your web creation process.

Drawing Project 16-1

Web Page

Create a CAD drawing web site. You can use either of the methods described in the chapter. Start by creating a home page. This will be the starting point and provide an index page for the CAD drawings. Save thumbnail drawing as .JPG files on this home page. From the .JPG drawings create hyperlinks to the .DWF pages. Create the pages by saving the drawings as .DWF files. Include hyperlinks by using the hyperlink command. Attach hyperlinks to any details on the drawings that you feel necessary. Turn your web site pages in to your instructor.

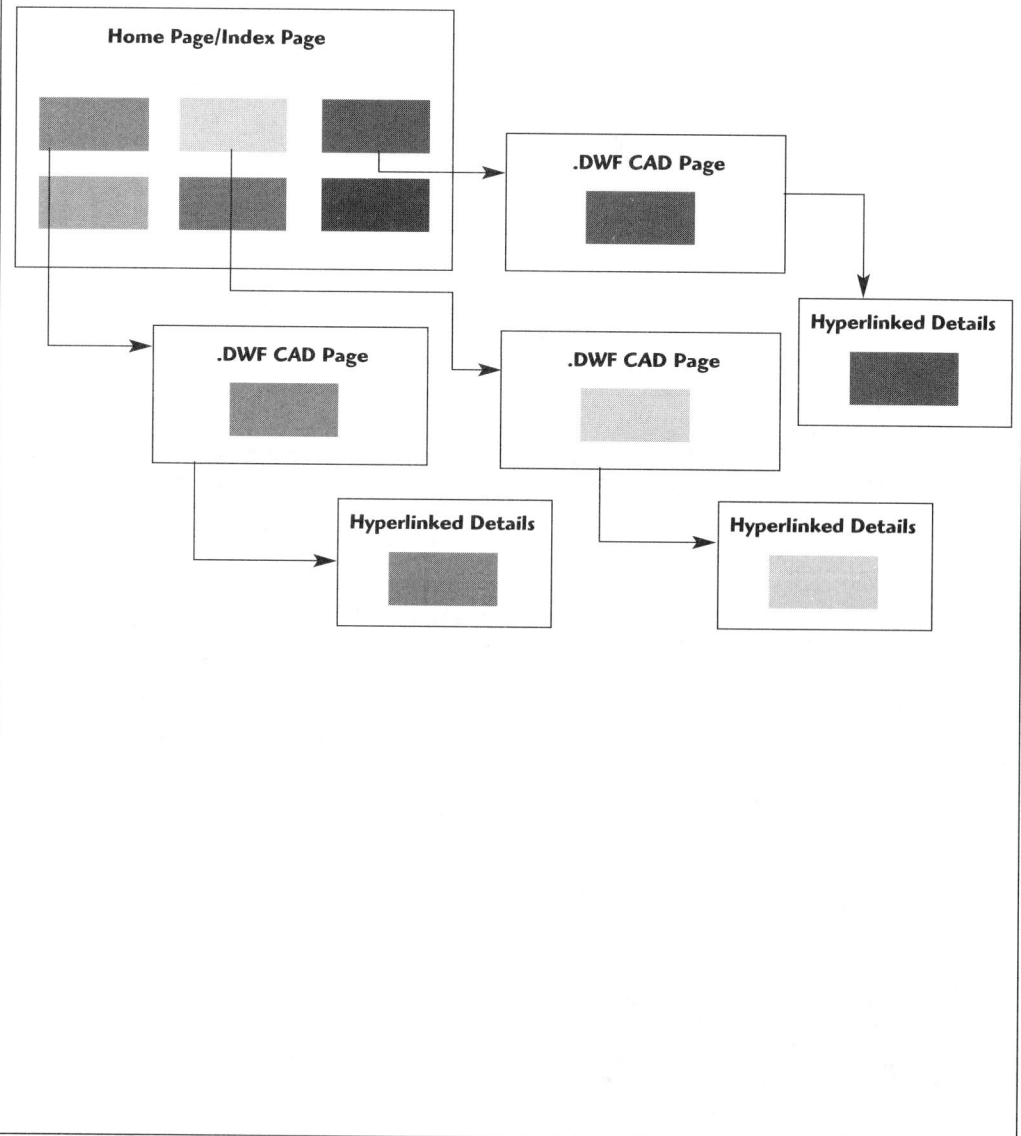

TOOLBAR QUICK FINDER

To bring up the available toolbars click:

View
> Toolbars:

Here are the common Mechanical Desktop toolbars and their locations:

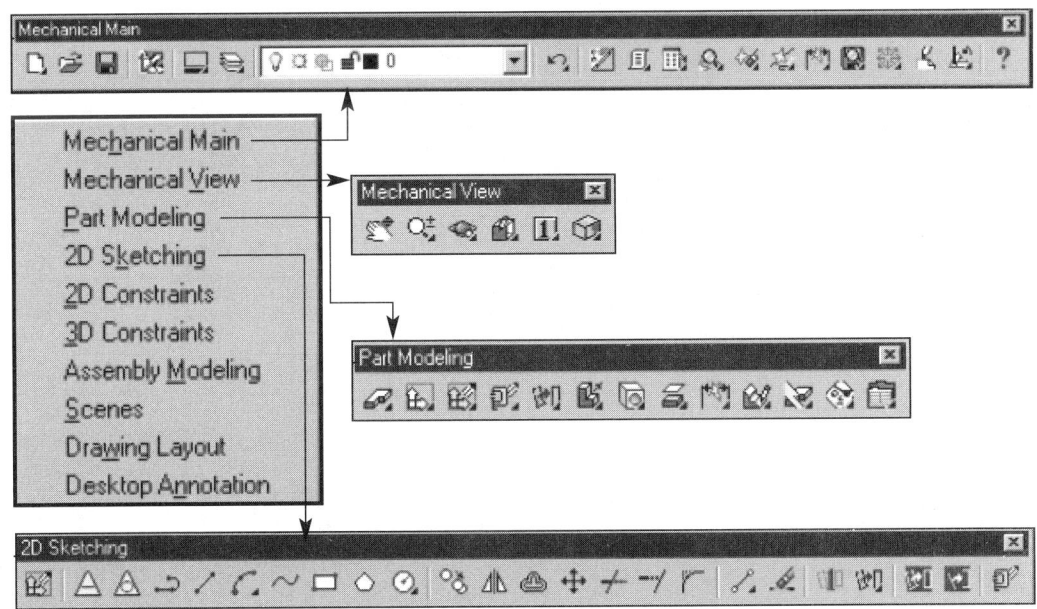

continued on next page

TOOLBAR QUICK FINDER

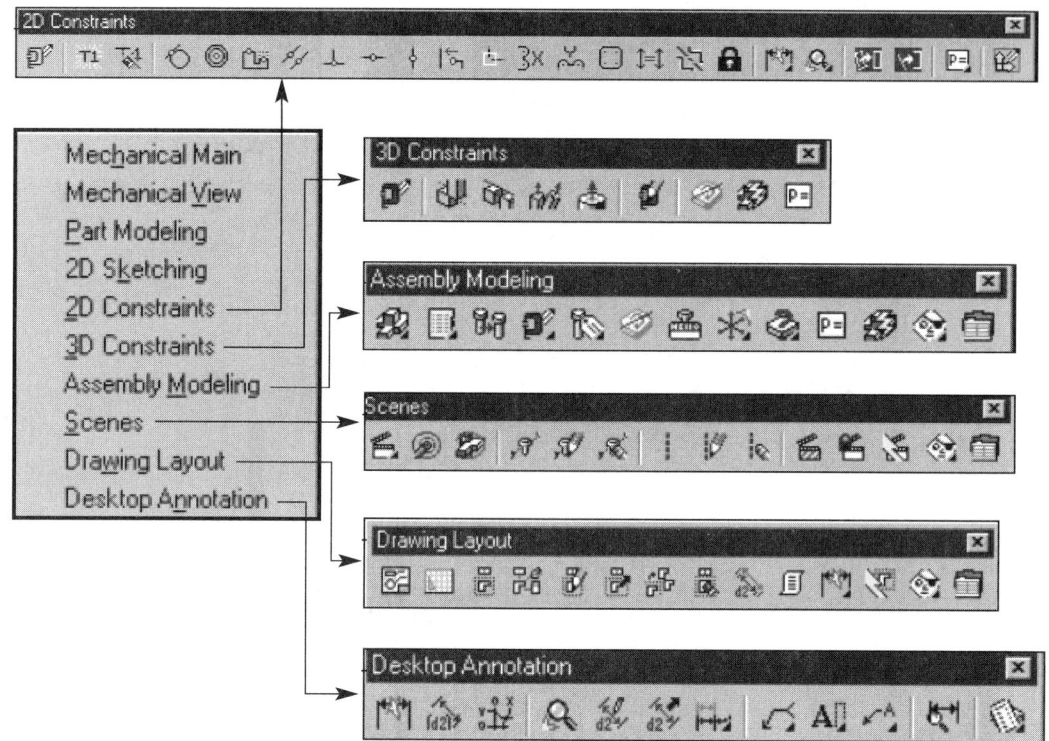

INDEX

A

Active Part, 110
Aligning (dimensions), 124
AM3DPATH, 143
AMACTIVATE, 184
AMADDCON, 23
AMBASICPLANES, 60, 141
AMBOM, 196-203
AMCATALOG, 50, 175
AMCOPYVIEW, 118
AMCOUNTB3D, 225
AMDELCON, 23
AMDELVIEW, 118
AMDIMALIGN, 125
AMDIMBEAK, 127
AMDIMINSERT, 126
AMDIMJOIN, 127
AMDIST, 193
AMDTPP, 40
AMDWGVIEW, 110, 121, 129, 183
AMEDIT, 201
AMEXTRUDE, 9, 14, 18, 53
AMFCFRAME, 212

AMFEA3D, 233
AMFILLET, 10
AMHOLE, 33
AMINSERT, 179
AMINTERFERE, 192
AMLISTVIEW, 118
AMLOFT, 88-96
AMMANIPULATE, 196
AMMASSMPROP, 189
AMMATE, 176
AMMODDIM, 12, 14, 17, 22, 56, 158
AMMOVEVIEW, 117
AMNEW, 49
AMNOTE, 122
AMOPTIONS, 120
AMPARDIM, 8, 13, 16, 20, 72
AMPARTLIST, 200
AMPROFILE, 8, 13, 16, 19, 71
AMPOWEREDIT, 123
AMREFDIM, 122
AMREVOLVE, 72-78
AMSCREW3D, 217
AMSHELL, 104
AMSHOWCON, 23
AMSKPLN, 51, 58
AMSTLSHAP3D, 220
AMSURFSYM, 207
AMSWEEP, 86-88
AMTRAIL, 181
AMUPDATE, 56
AMVIEW, 119
AMWELDSYM, 209
AMWORKAXIS, 58, 62, 142
AMWORKPLN, 59
Analysis, 236
ANSI (section patterns), 130
Assembly (s), 5, 173-92, 195
Assembly (toolbar), 195
Assembly Constraining, 176-82
   Align, 186, 187
   Flush, 186
   Insert, 186, 187
   Mate, 186

INDEX

Assembly Design tools, 189-93
Auxiliary Views, 133-35

B

Balloon Leaders, 202, 203
Base View, 110
Bill of Materials, 109, 196
Blind (extrusion), 28
Browse, 1

C

Chamfer, 36-37
Closed Cubic (lofting), 93
Close Profile, 12
Collinear Constraining, 24
Concentric:
    constraining, 24
    hole placement, 34
Constraining, 13-15, 17, 161
Constraint Toolbar, 23-24, 161
Copy (views), 118
Cubic:
    lofting, 93
    fillets, 37-38
Cut:
    extrusion, 30
    revolving, 77
Cutting Plane Line, 130-32

D

Desktop Browser, 54
Detail Drawings, 109
Detail Views, 132, 133
Dimension:
    as equations, 159, 160
    as numbers, 159, 160
    as parameters, 159, 160
Dimensions, 121-28
Draft angle (extrude), 30
Drawing Mode, 6-7

DWF (.DWF format), 243
Dynamically (Pan & Zoom), 29, 42

E

Edit (model), 54
Eplot (e-plot), 246
Equation (dimension), 20, 152
Explosion factor, 7
Extruding, 15, 23, 27-31

F

FEA, 231-39
FEA toolbar, 237
Feature, 14, 52
Fillets, 35-36, 156
Fix (constraining), 24, 159

G

Geometric Tolerance Symbols, 212, 213

H

Helix, 141
Holes, (AMHOLE), 32-34, 59, 155
Horizontal (Constraining), 24
Hyperlink (command), 247

I

Inserting (dimensions), 126
Interference (checking), 192
Internet tools, 243-51
Intersection:
    extrusion, 30
    revolving, 77
Iso View, 116

J

Join:
    extrusion, 30
    revolving, 76

L

Layout, (Views), 121
Linear:
    fillets, 37, 39
    lofting, 93
Lofting, 15, 88

M

Material (assign), 190
Math Operations, 159
Measuring (distance), 193
Mechanical View, 41
MidPlane:
    extrusion, 32
    revolving, 75
    shell, 106
Model Mode, 6
Modes of Display, 6
Moving (dimensions), 124
Mtext, 198
Multiple Thickness Override, 107

N

Normal to Start, 144

O

Offset, 66
On Edge/Axis, 61, 64
On Point, 59
On Vertices, 67
Orbit, 8, 29, 42
Ortho, 112-14

P

Parametric Model, 4, 10
Parametrics, 151, 158
    table driven, 161-67
Pitch, 143, 147, 148
Planar Angle, 65
Planar Normal, 64
Planar Parallel, 63
Polyline, 8
Power Pack, 217-28
Profiling, 14-15

R

Reclaim (shell), 106
Render (view), 29
Revolution, 148
Revolution Axis, 72
Revolving, 15, 71

S

Scene Mode, 6-7, 181
Screws (inserting), 226
Section Views, 129-32
Shaded View, 41, 43
Shaft Components (inserting), 223
Shell, 103
Sketch Plane, 50, 60
Solid Model, 3
Solved Sketch, 13
Spiral, 148
Split:
    extrusion, 31
    revolving, 78
Standardized Holes, 224
Steel Shapes (inserting), 220-22
Surfacing/Machining, Symbol, 207, 208
Sweeping, 15, 83, 146
Sweep Path, 83-85, 144, 145
Symbols Libraries, 2

T

Tangent (constraining), 24
Tap (holes), 33
Termination (blind hole), 33
Text (import), 199
Toolbar:
    constraints, 23
    mechanical view, 41

U

UCS, 57
Update Link, 165

V

Vertical (constraining), 24
Viewing Models, 40-44
Viewports, 44-45
Views (moving), 117

W

Web Page Creation, 243-51
Welding Symbols, 209-11
Wireframe Model, 3
Wireframe Views, 43
Wizards, 2
Work Axis, 58
Work Plane, 57, 60, 144
Work Point, 59
World (xy xz yz), 60

Z

Zoom views, 42

WHAT'S ON THE CD

The CD in the back of your book contains video clips to help you further understand some of the capabilities of Mechanical Desktop. The files are organized by chapter topics. The files are all executable files (.EXE), so to play them you need only to double-click on the file. You can use Windows Explorer to locate the chapter and topic you want, or click **Start,** then **Run,** and browse to the file to play.

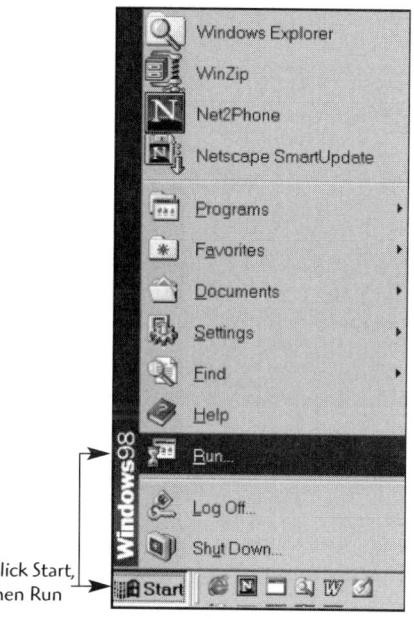

Click Start, then Run

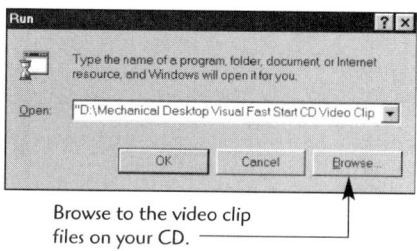

Browse to the video clip files on your CD.

The video clip will play, then return you to the previous screen.